THE DREAM UNIVERSE

ALSO BY DAVID LINDLEY

Uncertainty: Einstein, Heisenberg, Bohr,
and the Struggle for the Soul of Science

Degrees Kelvin: A Tale of Genius, Invention, and Tragedy

Boltzmann's Atom: The Great Debate
That Launched a Revolution in Physics

The Science of "Jurassic Park" and "The Lost World"
(with Rob DeSalle)

Where Does the Weirdness Go?: Why Quantum Mechanics Is Strange,
but Not as Strange as You Think

The End of Physics: The Myth of a Unified Theory

THE
DREAM
UNIVERSE

How Fundamental Physics Lost Its Way

DAVID LINDLEY

DOUBLEDAY   NEW YORK

In memory of Gwen Lindley, 1924–2011

Contents

Preface   xi

PART I HOW SCIENCE BEGAN   1

1 Galileo Invents Science   3

2 Copernicus Doesn't Quite Invent Astronomy   13

3 That Old-Time Philosophy   25

4 The Holy Roman Empire Strikes Back   34

5 How Science Uses Mathematics   49

PART II CLASSICAL SCIENCE REIGNS SUPREME   57

6 Mastery of Motion   59

7 The Language of Mathematics   71

8 The Limits of Pragmatism   82

PART III FUNDAMENTAL PHYSICS CHARTS ITS OWN COURSE   99

9 Dirac Invents Antimatter   101

10 Wigner's Enigmatic Question   110

11 All This Useless Beauty 122

12 Science and Engineering 134

PART IV SCIENCE OR PHILOSOPHY? 145

13 The Last Problems 147

14 The Byte-Sized Universe 162

15 Is Math All There Is? 172

16 The Dream Universe 182

Acknowledgments 203

Notes 205

Selected Bibliography 211

Index 213

Preface

Almost thirty years ago I published my first book, *The End of Physics*. You will no doubt be aware that there is still plenty of physics going on. But my focus in *The End of Physics* was more specific. I was discussing fundamental physics, the discipline that deals with the nature of matter at its most elementary level, the origin of the forces that hold matter together, and the formation of the universe itself. At the time, in the early 1990s, many scientists were all fired up about the possibility of building a "theory of everything"—a single, coherent intellectual framework that would capture all of fundamental physics in one neat and satisfying package. The point of my book was to say this ambition was delusional because such a theory could never be adequately tested. No telescopes or particle colliders, I said, would ever be powerful enough to see the finest internal details of the various proposed theories of everything, so that the connection between what the theories said and what we could actually observe and test was at best a lengthy and tenuous chain of inference.

I subtitled my book *The Myth of a Unified Theory*. By "myth" I meant that if physicists took a certain theory of everything to be the one true theory, it would not be because they could prove it right. It would be because they had collectively decided that it did a good enough job, that it was the best that could be reasonably achieved. But I was skeptical that even this lesser ambition could be realized:

it's impossible to imagine the community of theoretical physicists ever coming to the conclusion that they had done everything they could possibly do, that they had run out of questions to ask, and that it was time to close down the store and take up lawn bowls.

In one respect, what I said in *The End of Physics* proved correct. The very phrase "theory of everything" is little heard these days, and then mostly in a tone of knowing irony. The theorists who continue to ponder fundamental questions about our universe have largely embraced a very different but equally grandiose idea—that of the multiverse, an overarching conception that says that all kinds of possible universes exist in some parallel fashion, that our universe is one of many (and probably nothing very special to boot), and that what counts as fundamental physics will differ substantially from one universe to the next.

It was, no doubt, a great overreach to imagine that physics could ever account for every last jot and tittle about the construction and contents of our universe, but now the quest appears to have gone to the opposite extreme. According to the multiverse hypothesis, the answer to pretty much any question about how or why our universe looks the way it does is that there is no answer. In our universe, things look like this, but in some other universe they look like that. That's quite a comedown from the extravagant hopes of just a few decades ago.

And it raises a challenging question: What, exactly, are scholars of fundamental physics today trying to achieve? If their theories have no specific explanatory power when it comes to our universe in particular, then what larger question, if any, are they trying to answer? This new book, *The Dream Universe,* is my exploration of that question, and it comes to some conclusions that few physicists working on fundamental and cosmological issues will be happy to hear.

When I wrote *The End of Physics,* I had only recently emerged from my own professional immersion in research, working at Cambridge University and then at Fermilab on the connections between particle physics and cosmology. When I left academic research and decided to write about science rather than practice it, I became fascinated with the history of science. Since that time, I have written a number of books with an overtly historical perspective. My exploration of the scientific past has now come full circle and brought me back to my old subject: the state of modern cutting-edge fundamental physics. Reading and writing about history has taught me something that barely crossed my mind when I was practicing science—namely, that the idea of what a good scientific theory ought to look like is not set in stone, like some kind of ancient commandment. It is, instead, a matter that exists in the scientific atmosphere of its time; it is the common sense of its age, the unspoken ideal that scientists take for granted but think little about. And yet it changes profoundly from one era to the next.

That's why I chose to begin *The Dream Universe* with the story of Galileo Galilei, who laid the foundations for modern science by demonstrating that an astute combination of observation and reason could reveal how the natural world works. Galileo, as everyone knows, was condemned by the church for his heretical views, but in the first part of this book I dig into the true source of the church's opposition, which was not primarily theological but rather philosophical, reaching back to the thinking of the ancient Greeks. In doing so, I look at the Greeks' contribution to science: they were enamored of reason and logic, certainly, and they greatly esteemed the theorems of geometry, but for the most part they either ignored or positively disdained the sort of knowledge that could be gained by looking closely at the world they lived in. Measured by modern standards, what the Greeks espoused as an understanding of the universe at large was a form of idealistic philosophy very different from how we now conceive of scientific physics and cosmology. Only by breaking the hold of that ancient orthodoxy was Galileo able to think in a recognizably modern way. In particular, he used mathematics in a new way—not as a source of fundamental truths about the world but as a tool to manage the truths that observation and experiment delivered.

I delve into that notion in the second part of the book, which looks at how scientists of the classical era—up to the end of the nineteenth century, more or less—deployed mathematics with ever greater sophistication but thought of it always as a means of understanding the world, not as an arbiter or provider of fundamental truth.

That attitude began to change with the dawning of the twentieth century, as physicists sought to explore a world of fundamental phenomena increasingly removed from direct experience. In the third part of the book, I look at the rise of particle physics and modern cosmology and argue that while the power of mathematics has been essential in delivering deep theoretical understanding, it has also widened the gulf between theory and observation and, at times, led astray scientists who put too much faith in what mathematics tells them.

Those arguments set up the final part of the book, in which I assess some of the more extreme trends in fundamental physics and ask whether practitioners of this discipline can truly be regarded as scientists in the old Galilean sense. I conclude that in fact they have reverted to a premodern way of thinking about the universe in which, whether they acknowledge it or not, they expect that mathematics alone will provide all the answers. Fundamental physics has evolved into a version of philosophy—a highly mathematical form of philosophy, to be sure, but one that shares with other areas of philosophical inquiry an endless capacity to ask deep questions and an impressive inability ever to answer them.

HOW SCIENCE BEGAN

What we understand today as the scientific method of inquiry began during the Renaissance in Europe, and the individual most powerfully associated with its genesis is Galileo Galilei. The story is often told simplistically as one of a freethinking pioneer rebelling against the stodgy orthodoxy of the Catholic Church, but there's more to it than that. On matters of astronomy and mathematics, the church's orthodoxy was not particularly Christian but had its roots in certain branches of ancient Greek philosophy. Those philosophers are often credited with the earliest inklings of scientific thought, but that's an exaggeration. In their views on the nature of the universe the ancient Greeks evolved their own orthodoxy, and it was by breaking free of those constraints that Galileo found the path to modern science.

1

Galileo Invents Science

A book of geometrical problems published in 1561 included an analysis of how far a cannonball would travel, depending on the angle at which it was fired into the air. A charming illustration accompanied the discussion. It showed a cannonball flying from the mouth of the cannon in a perfectly straight line, then, having reached its maximum height, dropping straight down to the ground.[1] The relationship between the launch angle, the height, and the distance traveled was presented as an exercise in basic trigonometry—a problem any high school student of today would find straightforward but

a subject that was fairly new to would-be engineers, surveyors, and military planners of the sixteenth century.

What is bizarre to our modern eyes is not the use of trigonometry but the triangular path of the cannonball. The trajectory in the old book is an instance of Aristotelian dogma.

The ancients were perplexed by motion. It seemed to come in two kinds. Up in the sky, the sun and moon, the planets and stars all followed their predictable courses with unvarying steadiness. The gods, it seemed, had set the heavens in motion, and everything celestial would continue to move in the same way forever, with no further assistance. Heavenly motion was eternal.

Down on the ground, it was a different story. A stone kicked along the ground skitters on for a while, then comes to a halt. You can push a wheeled cart along a path, but unless you keep pushing, it will quickly come to a halt. Animals and people can move around, but they have to exert themselves to do so. For terrestrial objects, Aristotle decided, the natural state of affairs was not to be in motion but to be at rest. Motion only happened when something made it happen.

Except that wasn't quite right either. Drop a stone and it falls to the ground. A cart will roll downhill without assistance. That's because, according to Aristotle, all material objects are drawn to the center of the earth, which was in truth the center of the universe. Downward motion is the natural outcome of that tendency. (Weightless things, such as flames, are drawn to the heavens instead, and so go up.)

But that creates another problem. What is going on when you throw a stone into the air? It goes up and then comes down. Anyone could see that. But Aristotle said that motion only happens when something makes it happen. Once the stone has left your hand, it continues to rise. Something must still be pushing it, but what? This was a real puzzle for which no one found a very good answer. The most persuasive, if admittedly vague, solution was that when you throw the stone into the air, the action of your hand also has the effect of somehow raveling up the air in the vicinity of the stone; that air rises behind the stone, pushing it upward for a while until it unravels. And then the natural downward tendency takes over. What made the air rise behind the stone? Better not ask.

A version of this argument, at any rate, leads to the Aristotelian picture of how our cannonball would fly. It goes in a straight line, pushed along until the air gathered behind it runs out of oomph, and then it falls down.

The illustration reproduced here appeared in the book *Problematum astronomicorum et geometricorum* (Astronomical and Geometrical Problems) by Daniel Santbech, a Dutch mathematician. One might well think that even the most casual observer would have noticed that this is not how real cannonballs fly. They follow some sort of curving arc across the sky. Presumably people targeted by cannonballs would notice that they come at them laterally rather than plunging vertically down onto their heads, but perhaps they would have other matters on their minds at the time.

Here's the curious point, one that seems incomprehensible today: Daniel Santbech, like others of his age, was perfectly well aware that cannonballs do not follow triangular paths. If it bothered him to draw a diagram that clearly contradicted what anyone could see, he did his best to hide it. The purpose of the diagram was to convey the accepted orthodoxy of Aristotelian physics and match it to a simple trigonometrical calculation. Why did it appear that real cannonballs follow curved paths? That was a tricky matter, to be sure, but ancient philosophy prized reason and logic over everything else, including the evidence of the senses. Reason was the way to establish fundamental truths and inescapable conclusions. Humans and their senses were fallible and untrustworthy.

By the Middle Ages, Aristotelian philosophy had congealed into a fixed and unquestionable body of learning, preserved in ancient books that had come down from a wiser and more enlightened time. Much of this knowledge had been fused with the orthodoxy of the

Catholic Church, so that to question Aristotelian teaching, even on a seemingly undoctrinal matter such as the flight of a cannonball, was to run the risk of committing heresy. (It was up to the Inquisition, that fearsome instrument of the Roman Church, to figure out what was heresy and what was not, and their deliberations had a way of devolving into hairsplitting arcana impenetrable to outsiders.)

If there was an apparent conflict between orthodox teaching and observation of the world around you, your best bet was to stick with orthodoxy and conjure up some plausible reasons why one shouldn't necessarily believe what one saw. It was an awkward business, no doubt, but Aristotle wasn't around to sort things out, so his intellectual descendants and disciples had to wrestle with the task of figuring out how that most superior of intellects would have resolved the awkwardness.

The correct explanation of the flight of a cannonball appeared in 1638 in the pages of the *Discorsi e Dimostrazioni Matematiche intorno à due nuove scienze* (usually rendered in English as the *Dialogues Concerning Two New Sciences*), the last book written by Galileo Galilei. Born in Pisa in 1564, Galileo was then seventy-four years old, and living under house arrest in the small village of Arcetri, outside Florence, after his condemnation by the Catholic Church for embracing and promulgating the idea that the earth moves around the sun and not vice versa. The church had forbidden publication of anything Galileo wrote while under house arrest, but the *Dialogues,* after a lengthy struggle, saw the light of day in the Netherlands, where Protestantism had displaced Catholicism and the word of the pope was no longer feared. Or not feared as much as it had been, at least.

Galileo's troubles with the church sprang from his involvement in the fierce debate over Copernicanism, the hypothesis that the sun sat at the center of the universe, that the earth and other planets orbited around it, and that the moon went round the earth. For all the grief it brought him, astronomy was never Galileo's primary interest. His first insights, going back to when he was a student, had to do with the question of motion. He had experimented and theorized on motion his whole life, but it was only at the very end, shut up against his will in Arcetri, that he found time to compile his thoughts in a coherent manner and put them to paper. The "two new sciences" of the discourse were motion and the properties of solid objects (their strength, elasticity, and so on), and in both areas Galileo's thinking was revolutionary. But I want to concentrate on his innovations in motion and mechanics.

Galileo was the oldest of seven children, born to a father, Vincenzo, who was a professional musician and occasional music theorist, and a mother about whom little is known. The family moved from Pisa to Florence in 1574, when Galileo was ten years old. For a time, young Galileo received some education from the monks at Vallombrosa, a Benedictine monastery school about thirty kilometers from Florence. Something about monastic education appealed to him, and he became a novice, perhaps with the intention of taking holy orders. But his father, evidently a practical man, pulled him back in order to turn him into a student of medicine—that being, then as now, a reliable way for a bright young man to see his way to a prosperous future. Because he was the oldest son, the responsibility for taking care of the family after his parents were gone would fall largely on Galileo's shoulders. Dutifully, he enrolled in the University of Pisa in 1581 as a medical student.

Two years later, Galileo heard a lecture on Euclid by Ostilio Ricci, a mathematician in the service of the grand duke of Tuscany. Galileo was entranced by the clean intelligence and elegant logic of geometry. Here was a subject that delivered undeniable truths. Ricci was impressed by the young man's enthusiasm and introduced him to the work of Archimedes, one of the greatest mathematicians in human history, as well as a practical man who invented a variety of devices and machines. Ricci wrote to Vincenzo Galileo, imploring him to allow his son to switch his studies to mathematics, a subject for which he had evident love and genuine insight. But being a mathematician, in those days, was by no means a sure road to prosperity or respect. Mathematicians were largely disdained by the philosophers and theologians, the boss intellectuals of that era, as being little more than rude mechanicals, to borrow Shakespeare's phrase. Mathematicians might discover patterns in nature, but they would surely not find out underlying causes and reasons. Delving into the secrets of nature (or of God, which amounted to the same thing) was the province of philosophy. A few mathematicians, like Ricci himself, could make a living as part of a noble household—tutoring, surveying, assisting with navigational and military calculations, and so on. The rest might do the same kind of thing, but on a casual basis, picking up temporary work here and there as best they could. A sixteenth-century mathematician also might bring in a little cash by casting horoscopes. The study of the heavens, the geometrical paths of the sun and moon and planets, was seen as a department of mathematics, and astrology was one of its more saleable aspects.

In short, Vincenzo had no sympathy with his son's embrace of mathematics and insisted he stick with medicine, despite the pleas of the eminent Ricci. In the end, neither man won. Galileo, establishing a pattern that would mark his life, went his own way. He let his medical studies slide, went to lectures on mathematics and philosophy, and left the University of Pisa in 1585 without a degree in any subject.

For a few years, Galileo scraped along by teaching mathematics privately and casting horoscopes. But he was already pushing into uncharted intellectual territory. Modern scholarship reveals little of Galileo's personality—or rather, it reveals a variety of conflicting personalities. He was circumspect; he was reckless. He was modest; he was brash. He respected the church; he seethed at its strictures. All of these things may well have been true.

One story has it that, even as an undergraduate in Pisa, Galileo was apt to contradict his teachers.[2] The fundamental duty of a university teacher in those days, it should be said, was not to awaken a spirit of intellectual inquiry or to push the best students into the search for new knowledge, but to instruct pupils in the certainties conveyed in ancient books. A philosophy teacher would deliver Aristotelian wisdom, with any modern commentary designed only to patch up a few flaws and to do so strictly in the spirit of the original. A mathematics teacher would deliver Euclid along with heavenly geometry in the complicated system of Ptolemy, in which the earth sat at the center and the sun and the moon and planets revolved around our stationary home, traveling not in simple circles but in circles upon circles upon circles (they had to be circles, because philosophy demanded that the heavens be built from perfect geometrical forms, and circles were the most perfect of all).

Aristotelian physics held that heavier objects fall to the ground faster than light ones. Galileo, it is said, began to doubt this when it occurred to him that in a hailstorm, hailstones of all sizes, large and small, arrive en masse. That is hard to explain if the hailstones all descend from the same height. If any Aristotelian had thought about this, the response would have been dismissive. We know that heavier objects fall faster. That is the truth of the great thinkers of old. Mere observation of some sporadic phenomenon on earth is neither here nor there. Who knows where hailstones come from anyway?

By 1589 Galileo had managed to secure a position as a lecturer in mathematics at the University of Pisa. It was a poorly paid position, and the university itself was not among the best, but then Galileo's credentials were thin. Mainly, he had befriended a few notable people who were impressed by his knowledge and eagerness, and strings were pulled on his behalf.

Perhaps the most lasting tale about Galileo, apart from his later run-in with the Catholic authorities, is that he dropped objects from the leaning tower of Pisa in order to test whether, as he now believed, they would all fall at the same rate and take the same amount of time to hit the ground. The evidence that he did this amounts to later recollections plus hearsay from a number of people, and for a time modern historians doubted whether the experiments really took place. The weight of opinion now seems to be drifting back to the idea that he did indeed perform tests of this sort, but it is unclear when: as an undergraduate or later as a professor?

Many years later, in 1612, a skeptical philosopher at Pisa tried the same experiment, and reported that when he dropped objects of different weights from the same height, they didn't land precisely at the same time. In his *Dialogues Concerning Two New Sciences,* Galileo had a scathing answer to this petty objection. Aristotle had clearly stated in *On the Heavens* that "if a certain weight moves a certain distance in a certain time, a greater weight will move the same distance in a shorter time, and the proportion which the weights bear to one another, the times too will bear to one another."[3] In other words, Aristotelian principles demanded that a stone ten times heavier than another should hit the ground in one-tenth the time. The philosophy professor, Galileo continued, had dropped two objects of markedly different weight and found that the heavier one hit the ground a moment or two before the lighter one. And on this basis, Galileo asked, the professor wanted to claim that Aristotle was right and he, Galileo, was wrong?[4]

In essence, Galileo was admitting that an experiment such as his might not be perfect: he was aware of subtleties such as air resistance, as well as the practical difficulty of letting two objects go exactly simultaneously. The crucial point was that his proposition, that all objects fall at the same rate, was much closer to the truth than Aristotle's proposition that heavier objects fall faster. Galileo understood that getting closer to the truth was the important thing. This was a radical notion. Philosophers liked to deal with absolutes. Statements are true or they are not true. Galileo was saying that a little inexactitude was nothing to worry about. To the philosophers, this was an unsettling and dangerous suggestion.

By the time he wrote the *Dialogues on Two New Sciences,* Galileo had, after a number of false starts, figured out more precisely how objects fall. Downward velocity increases in proportion to the time spent falling, and the distance traveled consequently goes as the square of the elapsed time. Doing the experiments to establish these rules was no mean feat. Galileo had no stopwatch. Early on, he claimed that he counted time by his pulse (presumably he didn't get too excited when his experiments were going well). Later, he used a pendulum as a timer. Figuring out that a weight on a string, swinging back and forth, keeps constant time regardless of how far it swung back and forth was another notable achievement (this is not precisely true, as it turns out, but it's close enough that pendulum clocks were a mainstay of timekeeping for centuries).

Over the years, Galileo had also come to question and ultimately reject another Aristotelian principle—that motion in the terrestrial sphere, unlike motion in the heavens, demanded continual impetus to keep it going. He had, in his later years, fully accepted the idea of Nicolaus Copernicus that the sun was at the center of the universe and that the earth and other planets went endlessly around it, and that the moon likewise went around the earth. Why should the same rule not apply on earth itself? What stops a cannonball from flying off forever? Because it falls to the ground, and indeed begins to fall, as Galileo eventually realized, as soon as it leaves the mouth of the cannon. The two motions—along the line of flight and down toward the ground—happen together. It's not the case, as the Aristotelians believed, that you get one and then the other.

There was another point of contention between the old way of thinking and the new. In Aristotelian philosophy, objects fell toward the center of the earth because they sought out the lowest place. To our modern way of thinking, this sounds like no explanation at all. It seems like a perfect tautology: things fall because it is their fate to fall. But to the committed Aristotelian it made sense—it provided an explanation rooted in the idea of purpose. In the Aristotelian world, things happened in order for some end state to be achieved. The philosopher's task was to figure out those purposes, no matter how specious the reasoning might seem to us now.

Galileo, by contrast, said nothing about *why* objects fall. His method was to observe closely *how* they fall, and thereby construct an accurate quantitative description. It was not that he was incurious about why objects fell to earth, but that he understood it to be a question he could postpone without detracting from his ability to make a mathematical statement about the way things fell. To the Aristotelian mind, this was pointless: quantitative analysis was seen as window dressing at best. What was the use of a mere description of events without any cause being imputed? Where, in other words, was the intellectual heft of what Galileo was doing?

We see things differently now. Galileo had the right idea: bite off what you can chew, and don't pretend to know things you don't. Armed with his ideas about motion, it was now a relatively simple matter for Galileo, using nothing more than classical geometry, to figure out what the path of a cannonball must be. The answer was a parabola, a curve known to the ancients from their studies of geometry. It was an imperfect sort of curve, if you believed that only circles and parts of circles bore the divine fingerprint, but Galileo found a place for it in the real world. He had taken a commonplace worldly phenomenon—the flight of a cannonball—analyzed it in the light of his lifelong observing, theorizing, and testing of the rules of motion and had determined the mathematical form of the cannonball's trajectory. He had proved it was a parabola. He had used observational and experimental information to infer a reliable mathematical rule.

He had invented science.

2

Copernicus Doesn't Quite Invent Astronomy

I t's sometimes said that the first modern science was astronomy. Nicolaus Copernicus's *De revolutionibus orbium coelestium* (*On the Revolutions of the Heavenly Spheres*), which presented the startling idea that the sun was at the center of the universe and that the earth moved around it, appeared in 1543, just before Copernicus died at the age of seventy, and twenty-one years before Galileo was born. But rumors of the heliocentric system and the Polish cleric behind it had been circulating among the astronomers and mathematicians of Europe for at least thirty years before that. By the time Galileo had begun his life's work, heliocentrism was the subject of impassioned debate, although its full acceptance had to wait until after his death.

So astronomy indeed got a head start on physics. But although Copernicus upended the old heavens by putting the sun at the center and the earth to one side, he was a traditionalist in other respects. *De revolutionibus* was a work of geometry. There was no physics in the Ptolemaic heavens, and there was none in the Copernican heavens either. Copernicus's goal was depiction only; his aim was to devise a geometrical system that reproduced the motion of the sun, moon, and planets across the sky, and which could be used for navigation and for creating calendars that put Easter and other holidays at the right time of the year (Copernicus was a canon at the

Catholic cathedral in Frombork, in what was then Prussia and is now Poland). The old Ptolemaic system, which had the earth at the center with all other bodies revolving around it, was accurate, but it was fantastically complicated and correspondingly awkward as a tool for calculation. Putting the sun at the center made the geometry easier. Many astronomers and mathematicians, especially those in Protestant countries, eagerly adopted Copernicus's new system because it greatly simplified calendrical calculations.

But it was still not all that simple. Although Copernicus cast aside one old philosophical decree by permitting the earth to move, he stuck with another, which was that all the orbits in his system had to be perfect circles. The philosophical argument was this: motion in a circle, constant and unending, could be seen as divine handiwork, instigated long ago and continuing ever after. Other forms of motion—a path deviating from strict circularity, a speed going up and down—surely required the intervention of some agent to cause the changes. God would assuredly not create an imperfect heavenly system that required the services of a janitor to maintain it.

Constructing a system of circles that precisely reproduced the paths of the planets was no easy feat, and the difficulty of the task is part of the explanation for Copernicus's reluctance to publish: he was never sure he had got it quite right. As the Harvard astronomy historian Owen Gingerich describes it, *De revolutionibus* was about 5 percent descriptive preface, laying out the heliocentric system, with the remaining 95 percent a "deadly technical" account of geometry and calculation and tabulation off-putting to all but the most valiant readers.[1]

The emergence of heliocentrism is an oft-told story, and a brief summary will suffice. Accurate observation of the skies goes back to the Babylonians. In the second millennium BCE, they plotted the tracks of the sun, moon, and visible planets (Mercury, Venus, Mars, Jupiter, and Saturn) across the sky, and learned enough to predict solar and lunar eclipses. Their chief interest was astrology, but correctly interpreting the will of the gods requires precision, and they got a lot of things right. Their knowledge of the heavens was absorbed into the Western tradition via Egypt, Greece, and Rome, and survived largely because of Claudius Ptolemy, who lived from around 100 to 170 CE in Alexandria, Egypt, then under Roman rule. He wrote in Greek, but his great work on astronomy survived the fall of Rome through its preservation in the hands of Arabic scholars and is known best by its Arabic name, *Almagest*. Identifying all the sources that Ptolemy stirred into *Almagest* is an impossible task. What came down to Europe in the Middle Ages as part of the whole package of Aristotelian natural philosophy is known, nonetheless, as the Ptolemaic astronomical system.

With the earth at the center of the universe, the firmament of stars wheel around as if they were lights attached to the inner surface of a distant sphere revolving in stately fashion. This was straightforward. What gets fiddly is explaining the paths of the handful of objects that move differently from the stars—sun, moon, and five planets—on

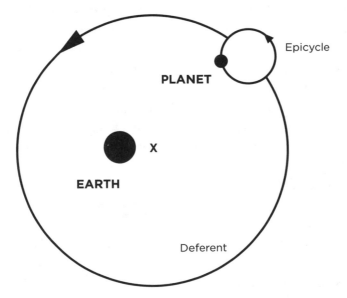

the assumption that they also revolve around the earth, within the sphere of the stars, and move strictly on circular paths.

As seen from earth, Mars, for example, follows a complicated path. Its speed across the sky is highly variable and it sometimes reverses direction altogether, in what is called retrograde motion. Ptolemy's first trick was the epicycle. He put Mars on a second circle centered on the circumference of another, giving the planet a trajectory that intermittently loops back and forth. But that still wasn't the whole story. Philosophy required that heavenly paths had to be circles, and that bodies carried by those circles had to move at constant speed. Those restrictions made it impossible to reproduce the path of Mars with an epicycle alone, so Ptolemy created a little more flexibility by putting the earth slightly apart from the center of the circle, and declaring that planetary motion was constant as seen from this offset position. Ptolemy called this displacement an "equant,"[2] and it enabled him to get the motion of Mars to match up better with the observations. With epicycles and equants for each planet, and a handful of other ruses, the Ptolemaic system was born. It was a Rube Goldberg contraption, to be sure, but since the idea was not to explain the construction of the heavens in any sense that we would now accept as scientific, only to provide a means for calculating planetary positions, it served its purpose.

Even in the Middle Ages, however, Ptolemaic astronomy had its detractors. The equant, in particular, was widely regarded as a cheat, a geometrical gimmick that obeyed the letter but not the spirit of Aristotelian thinking. But at the time, no one could come up with anything better.

Then came Nicolaus Copernicus, born in 1473 in Toruń, Poland, into a wealthy and well-connected (on his mother's side) family. After he was educated at the University of Kraków, family influence steered him into a position as a canon in the Catholic region of Warmia on the Baltic coast of what is now northern Poland. Before taking up that position permanently, Copernicus spent time at the University of Bologna, the oldest in Europe, and also at Padua and Rome. He delved deeper into astronomy, encountering criticisms of the Ptolemaic system for its inaccuracies in certain fine points of celestial motion. He returned to Warmia for good in 1503 and in about 1510 became a canon in the cathedral city of Frombork. He spent the rest of his life there, employed by the church but toiling over astronomy.

The idea of putting the sun at the center of the universe was not entirely new. The Greek mathematician Aristarchus, active in the third century BCE, had suggested not only that the earth moves around the sun but also that the earth rotates on its own axis. The common objection, then and for many centuries afterward, was that the earth could not possibly be moving, still less rotating, because everything on the surface would fly off at great speed. For almost all philosophers, this was an insuperable objection.

Even so, Copernicus could see that having the earth and other planets revolve around the sun conferred a great simplification: it did away with the need for Ptolemy's large epicycles. Sometime around or before 1514, Copernicus wrote a short pamphlet, the *Commentariolus,* or *Little Commentary,* that sketched out how a sun-centered system would work. This pamphlet was never formally published but circulated around Europe and began to be whispered about by other mathematicians and astronomers. For centuries it existed only as rumor and hearsay, but in the late nineteenth century a copy was found in Stockholm. A few others have shown up since then, including one with a library record dated 1514.[3]

The *Little Commentary* laid out the essential problem Copernicus had set himself. Making the planets go around the sun was a simplifying stroke. Insisting that their orbits be circular made everything difficult again. To get his model to work, Copernicus resorted to epicyclets, or little epicycles. In effect, he got rid of the big epicycles of the Ptolemaic system and replaced the much derided equants with little epicycles, small circles adorning the planets' orbits that accounted for their varying speeds across the sky.

This is a gross oversimplification but it conveys the essential point: the Copernican system was simpler than the Ptolemaic system, but it was not simple. The story of how *De revolutionibus* came into the world is fascinating but lengthy. Owen Gingerich, in *The Book Nobody Read,* lays out the tale. In brief, a young scholar made it his task to ensure Copernicus's work became public, went to Frombork to encourage and badger the increasingly frail old man to finish his calculations or concede that they were as good as they were going to get, made arrangements with a publisher in Nuremberg to print the book, knowing the controversy it was likely to cause, and personally delivered the manuscript to the printer in 1543. It's unclear if Copernicus ever saw a finished copy of his own book. An apocryphal story has him putting his hand on a copy and smiling contentedly before passing into the great beyond, but this is surely too cloying to be true.

In its first several decades, *De revolutionibus* circulated among the intellectuals of Europe without earning any great hostility from the Catholic Church. The book came with a preface, assumed to have been written by Copernicus, that declared in no uncertain terms that the heliocentric system was to be regarded purely as a mathematical hypothesis, a geometrical recipe for figuring out the positions of the heavenly bodies. An astronomer, even a Catholic astronomer, could draw on the mathematical virtues of *De revolutionibus* without making any commitment as to its truth—the latter being an issue best left to theologians.

The first indication that Galileo agreed with the literal truth of Copernicanism came in 1596, when he was thirty-two years old. He had received a copy of a new book, *Mysterium cosmographicum* (*Mysteries of the Cosmos*), written by Johannes Kepler, a German mathematician. Kepler made plain his full agreement with heliocentrism, and Galileo wrote a short note of thanks to him, acknowledging that he, too, knew of many arguments in favor of putting the sun at the center of the universe (including some, he said cryptically, of his own invention), but that in the intellectual climate of Italy he was fearful of saying so in public. In his reply, Kepler admonished him to make his position known. All the top mathematicians, Kepler said, were heliocentrists now. But Kepler lived in a Protestant country and Galileo was in Padua, where the Vatican held sway. There is no sign that Galileo replied to Kepler's letter.

Kepler was an odd character, a combination of old-fashioned mystic and forward-looking astronomer. Born in southwestern Germany, he showed early ability in mathematics and learned both the Ptolemaic and the Copernican versions of astronomy. He soon decided he preferred the latter. He became a teacher at a Protestant school in Graz, Austria, and devoted himself to updating and improving Copernicus—in the service, however, of a far more ambitious cosmological scheme. His *Mysterium cosmographicum* reflects both aspects of his nature. The book offered technical refinements of Copernicus's system, but the bulk of it was devoted to a grand system that, Kepler believed, revealed the divine plan for the universe.

Kepler jettisoned various assumptions of Aristotelian philosophy but replaced them with ideas that had even older roots. He pictured the solar system as being built from a collection of three-dimensional geometrical shapes, the regular Platonic solids, nested one inside the other. The Platonic solids are the five objects made of equal two-dimensional shapes joined at their edges: the tetrahedron (four equilateral triangles making a pyramid); the cube (six square faces); the octahedron (eight equilateral triangles); the dodecahedron (twelve pentagonal, or five-sided, faces); and the icosahedron (twenty equilateral triangles). Plato was greatly enamored of these shapes, because of their regularity and because there are only five of them. That was enough for him to assert in the *Timaeus* that they must be the fundamental shapes of the elements that comprise the earth. Sadly, classical Greek philosophy allowed for only four elements: earth, air, fire, and water. Happily, Aristotle later came up with a fifth element, quintessence, which was supposed to fill the superlunary space beyond the moon and which was therefore never to be found on earth.

An arrangement of nested Platonic solids, Kepler said, would determine the relative sizes of the orbits of Mercury, Venus, Earth, Mars, Jupiter, and Saturn, each planet residing in some way on a shell of its assigned geometrical shape. It would have been very pleasing for Kepler if he had been able to do this by nesting the solids in their natural order: tetrahedron, cube, octahedron, and so on. But this didn't work, and after much effort he found a sequence that worked to his satisfaction: octahedron, icosahedron, dodecahedron, tetrahedron, cube. The tidy arrangement of these beautiful shapes in an order that gave the correct sizes of the planetary orbits was a sure indication, in Kepler's mind, that he had uncovered God's cosmic architecture. But as Nobel laureate Steven Weinberg has pointed out, there are many ways to order the five solids, and given that Kepler's model is not totally accurate anyway, it's not so surprising that Kepler was able to find a sequence that worked well enough.[4]

It's highly unlikely that Galileo gave *Mysterium cosmographicum* his close attention. Galileo was not much interested in celestial motion at that time, and the book's cosmic plan would have repelled him. Gushing about the grandeur of Platonic geometry and how it must hold the essential truths of a new cosmology was exactly opposite to Galileo's way of thinking. His principle was to make accurate observations and measurements, reason carefully about their meaning, and thus find the right mathematics to fit what he discovered. Mathematics, as Galileo understood it, was the means to capture physical ideas in quantitative form, but it was not, as Kepler wanted to believe, the *source* of those ideas.

Other events soon forced Galileo to take greater interest in astronomy, however. In October 1604, a new star—*nova,* in Latin—appeared in the night sky. It became brighter than any other star, and was visible during the day for three weeks. Kepler was among those who observed the strange phenomenon, and he summarized his views in his book of 1606, *De stella nova.* Two facts stood out. First, the new star maintained its position among the fixed stars of the sky, unlike the moon and planets, which move daily across the stellar background. The nova must therefore be farther away than any planet. Second, it was transient: it appeared out of nowhere, was prominent for a time, then over a period of months faded away.

Aristotelian astronomy held that everything beyond the moon is unchanging: the stars shine forever, the planets follow their prescribed courses in perpetuity. The existence of a transient event in the superlunary realm was philosophically untenable. But there it was.

Galileo, asked to present a lecture in Padua on the astonishing celestial sight, could only agree: whatever the new star was, it was a temporary phenomenon far beyond the moon. Some defenders of the old philosophy resorted to a familiar strategy: when your theory is in doubt, attack the observations. Such critics claimed that the new star (we now know it as a supernova, a huge stellar explosion somewhere in our galaxy) did in fact move with respect to the fixed stars, and said it was a comet, traveling in the variable realm within the moon's orbit and therefore philosophically acceptable. But observations by reputable astronomers everywhere refuted this argument.

What drew Galileo unapologetically into the camp of Copernicus and Kepler was the invention of the telescope in the Netherlands in 1608. Galileo learned to make his own, better than those he could obtain from others. He studied Jupiter, and found that it had four little lights clustered around it. Following them over several months, he concluded that they were moons of the planet, orbiting Jupiter just as Copernicus said the planets orbited the sun. Not proof of Copernicanism, exactly, but a powerful hint.

The more telling evidence came from Galileo's observations of Venus. He found that it showed phases, just like our moon, waxing from thin crescent to full and back again. This was exactly what the Copernican model predicted, with Venus going from full when it was on the other side of the sun and, to an observer on earth, fully illuminated, to a bare sliver when Venus was between earth and sun. Venus also had phases in the Ptolemaic system, but they were much less pronounced. What's more, the Copernican system explained another puzzle, which was that the brightness of Venus didn't vary all that much despite its phase: the reason is that when Venus is full it is farthest from earth, and when it is a crescent, reflecting less of the sun's light, it is much closer.

This argument grabbed Galileo's attention. It depended on specific and calculable quantities, derived from a mathematical model, that could be compared with what an observer saw. The four major moons of Jupiter (the Galilean moons, as they are now called) were an example of a Copernican system in miniature. But the explanation of the phases and brightness of Venus was a verifiable solution to a curious problem. Copernicus's model of the solar system wasn't mere geometrical hypothesis—it was a picture of the solar system as it actually is.

Galileo quickly wrote up his work with the telescope and published in 1610 an enormously influential book, *Sidereus nuncius,* usually rendered in English as *The Starry Messenger,* although it might also be translated as "A Message from the Stars." A certain ambiguity arises: Was Galileo representing the telescope as a messenger bringing news about the heavens or was he casting himself as the messenger? He was likely very aware of the inherent ambiguity in his phrasing, which allowed him to declare himself the bringer of scientific news but at the same time left him room to stand back, should the message prove controversial.[5]

Controversy indeed came. Galileo's magnificent summation of what he had learned from his telescopic observations made him famous across Europe. His earlier investigations of falling bodies and the swinging of pendulums were remarkable but were not matters likely to excite great public interest. Rearranging the heavens, on the other hand, and carefully listing his reasons for doing so—that was extraordinary.

The old-school philosophers fought back. One, hearing about the novel celestial sights that the telescope revealed, simply refused to look. Not all telescopes were of the necessary optical quality or magnifying power, so some people looked and saw no Jovian moons or failed to clearly discern the phases of Venus. Other philosophers raised a more subtle objection: How could anyone assert with confidence that a telescope, tested and proved only in the terrestrial sphere, would work the same way on heavenly bodies?

Kepler was also pushing ahead. In 1609, he published a new work, *Astronomia nova* (*The New Astronomy*). This volume recounted his painstaking analysis of planetary observations compiled by the great Danish astronomer Tycho Brahe. After a contentious first encounter, Brahe had agreed to employ Kepler as his assistant, but died not long afterward, in 1601. Kepler was then appointed to Brahe's position in Prague as imperial mathematician to Archduke Ferdinand of the Holy Roman Empire, with the chief task of revising astronomical tables for calendrical purposes. As he pored over the more accurate planetary positions to which he now had access, he found it increasingly difficult to get Copernicus's system to work precisely. Adjustments would resolve discrepancies in one place, only to generate new errors elsewhere.

Eventually, Kepler hit on a much better solution. He found that he could marry Brahe's observations of the position of Mars in the sky to a model in which its orbit around the sun was not composed of multiple circles but was a single ellipse, with no need for epicycles or other adornments. Just as Galileo had used the parabola to model the flight of a cannonball, now Kepler used another imperfect shape, the ellipse, to depict the path of Mars. And he could explain the orbits of the other planets with their own ellipses.

This was another shock to the Aristotelians. Their carefully ordered cosmos was being dismantled and replaced with mathematical constructions built from philosophically unworthy shapes.

Kepler has been called the first astrophysicist. His decision to try elliptical orbits came not merely from the urge to tinker with celestial geometry but because he was thinking about how and why the planets move as they do. The sun, to Kepler, was the motive power, the engine of the universe. He knew that Mercury, closest to the sun, moved the fastest among the planets, that Venus, farther away, moved more slowly, and so on. Therefore, he reasoned, if a planet were to move a little closer to the sun, it would go faster, and conversely orbit slower as it retreated. This is exactly what his elliptical orbits accomplished: they matched not only the shape of the planets' paths across the sky but also their speeds along those paths.

But was this real physical insight or a matter of being right for the wrong reason? Kepler's idea that planets move faster the closer they are to the sun's power seems perilously close to the Aristotelian principle that objects fall to earth because they are drawn to the center of the universe. Kepler's undoubted mathematical skill was allied in his mind with a fundamentally mystical view of the heavens, so to suggest that his invocation of elliptical orbits rather than circular ones derived from an insight we could call physics seems a reach, at least to me.

Still, Kepler's innovation, however it came about, was a huge step forward. He argued ever more forcefully for the literal truth of heliocentrism and, necessarily, for the idea that the earth must be moving. Galileo, more cautiously, leaned in the same direction, and his arguments brought, as he feared they must, unwelcome attention from the church. Ostensibly, the dispute between Galileo and the church came down to a single point: Did the earth move through space or was it, as Catholic doctrine required, stationary? But at the heart of the argument was a much larger question, whether truth was to be found in old books or instead established by investigation and reason. More than anything, what came between Galileo and Rome was an argument over how truth was to be known and who could be entrusted with it.

3

That Old-Time Philosophy

The one-sentence account of Galileo's strife with the Catholic Church is that he got into trouble when he began publicly airing his opinion that the earth moved around the sun, an assertion that Rome found heretical. That's true, but it raises a host of questions. Why did the church find the relationship of sun and earth a matter of doctrinal significance in the first place? And how, more broadly, did Catholic orthodoxy come to be identified with Aristotelianism on a whole slew of matters over which the Bible itself was silent?

Among scientists today—or more accurately, among those who engage in what are usually called the hard sciences, physics, astronomy, chemistry, and the like—Aristotle is the chief villain among ancient philosophers. This opinion is standard for those scientists who have read a little history and philosophy and is probably believed even more vehemently by those who have read neither but have seen references here and there to the pernicious hold of Aristotelian thinking in the prescientific world. But Aristotelianism, as practiced in pre-Renaissance Europe, was a highly refracted interpretation of Aristotle's principles and opinions, and embodied an intellectual dogmatism that was uncharacteristic of the man himself.

Aristotle was prolific and wide-ranging. He wrote on politics, on morals and ethics, on drama and rhetoric. His virtues and flaws on these matters are not our present concern. In his book *The Lagoon:*

How Aristotle Invented Science, evolutionary biologist Armand Leroi attempts to rehabilitate Aristotle by positioning him as a true early scientist. Leroi's concern is biology, especially zoology, and in that area Aristotle undertook his own investigations and came up with original opinions. He was fascinated by the complexity and variety of the animal life he saw around him and strove to make sense of it. He was a classifier and cataloger. He described creatures at great length, dissected them to understand differences and similarities in their internal organs, divided those that laid eggs from those that produced live young and those that generated larvae, distinguished those with fur or hair from those with bare skin, made distinctions between fish with gills and sea creatures that had blowholes and, he correctly deduced, lungs, and so on and so forth.

This kind of extravagantly detailed analysis was not at all what other philosophers—notably Plato—undertook. Plato, in fact, positively disdained empirical investigation, saying that it detracted from the proper pursuit of philosophy, which was to figure out the nature of the universe through logic and reason. But Aristotle used reason in his own fashion. He believed firmly that the variety of animal life had its own logic. "Nature makes nothing in vain" was his mantra. As he explored and documented the zoological world, he sought reasons for the way animals were, why they were made in such a manner, and how their form and structure were appropriate to their way of life.

This sounds very fine, and Aristotle asked many good questions. Ultimately, though, he had no way of answering them. Though he did his best to devise explanations, the strategy he resorted to most often was the decidedly nonscientific one of making stuff up. Hedgehogs, he said, copulate by standing chest to chest on their back legs, to avoid sticking each other with their spines. Elephants (which he had not seen, only read about in reports from others) live in swamps, to take the weight off their stubby little legs, and breathe through their trunks. Oysters come to life through spontaneous generation:

Aristotle had seen them appear in the bottoms of pots thrown overboard from ships; since oysters cannot move, they must have sprung up from silt and mud in the pots.[1]

Picking on these examples is unfair, to be sure. Aristotle was no fool, and his observations and inferences were often apt. Leroi's case is that he had the probing mind of a scientist and the insistence on searching for reasons, and that in these ways he foreshadowed true scientific investigation. The weakness of the argument is that he didn't really know—nobody did, after all—how to go about a scientific investigation, how to develop ideas and hypotheses and put them to the test. Rather, what he did was come up with plausible-sounding explanations and sift through his voluminous catalog of facts for examples that fit them. Cherry-picking, we would call it now.

The Aristotelian style has not entirely disappeared. There was and perhaps still is an Internet-driven theory that the increasing rate of breast cancer in recent decades is tied to the use of underarm deodorants. By preventing sweating, deodorants also prevent certain toxins from escaping the body; trapped inside, they migrate to breast tissue and trigger cancer. The internal logic of this explanation has a certain appeal. What's lacking is any evidence for the specifics: What are these toxins? Does sweating in fact expel them? Why do they cause breast cancer rather than skin cancer or bone cancer?

The big problem with Aristotle, one scholar has said, is that "his theories are hasty generalisations based on rather superficial observations."[2] On physics and astronomy, moreover, Aristotle was much less impressive. He was a qualitative thinker. He sought cause and effect, but his criterion was plausibility, not quantitative precision. He was not a mathematician. What he wrote on astronomy, as he clearly acknowledged, derived almost entirely from his predecessors. His system of the heavens came from Eudoxus, who was inspired by Plato, who took his cues from Pythagoras, or rather from the cultish fifth-century BCE group who became known as the Pythagoreans.

Pythagoras was born around 570 BCE on Samos, an island in the eastern Aegean, but moved as a young man to Croton, on the southern coast of Italy, where a group of fellow philosophers and acolytes gathered around him. The famous Pythagorean theorem about triangles was certainly not his invention—the Babylonians knew it—nor is it clear, as some have suggested, that he (or one of his followers) provided the first rigorous proof. More than anything, Pythagoras appears to have been an ascetic figure whose teachings about a pure lifestyle were the foundation of a movement that came to have political influence in the Greek world. Pythagoreanism entailed a good deal of mysticism, an important part of which revolved around numbers and their significance. Three was male and two was female. Four represented justice and seven stood for opportunity. The Pythagoreans also believed, it seems, that numbers were literally the stuff of the universe, the entities from which everything else was made. What this meant was far from clear. Much of this teaching and the association of specific numbers with specific ideas was, as G. E. R. Lloyd has put it, "quite fantastic and arbitrary."[3]

Still, there were some real achievements. Pythagoras (or, again, perhaps one or several of his followers) discovered that the lengths of plucked strings are related to the musical notes they sound, and that combinations of notes that please the ear come from strings whose lengths are related by simple ratios: 3 to 2, 4 to 3, etc. This neat alignment of numerical ratios and musical harmonies was the jumping-off point for a cosmic scheme of harmoniousness. The Pythagoreans said that the sun, moon, and planets were set at distances related by simple arithmetical ratios. There was literal harmony in this arrangement: the planets emit sounds as they travel through the heavens, but we, having been bathed in this celestial music all our lives, cannot hear it. Later writers elevated this charming idea into a grand system of cosmic design, the "harmony of the spheres." Kepler's system of Platonic solids is one of the last appearances of this principle among respectable thinkers.

What the Pythagoreans said about the structure of the heavens is a confusing mess. Some put a cosmic fire at the center of the earth, with the earth at the center of everything else. Others said the earth was not noble enough to be central. They put the fire at the center and the earth to one side, with a supposed "counter-earth" between us and the fire to explain why we can't see it (the sun was not this fire but a separate body). In this scheme, the total number of heavenly bodies—earth, counter-earth, sun and moon, five planets, and the sphere of the stars—came to ten, which was deemed to be a very good thing.

The specifics of these cosmic architectures had little lasting impact. What was far more influential was the insistence that cosmic architecture must be built on elegant mathematical relationships. Our word "cosmos" derives from the Greek *kosmos,* meaning an orderly or harmonious arrangement. It's the opposite of *kaos,* the void or the abyss. One way or another, this principle of harmony has twisted and turned through the history of philosophy and then science, and survives today. The man who gave this idea lasting respectability, detaching it from the mystical woo-woo of the Pythagoreans, was Plato.

Like Aristotle, Plato has a complicated relationship with modern science. By no means a mathematician himself, he was greatly enamored (too enamored, perhaps) of the Pythagorean enthusiasm for mathematics, especially by the notion that mathematical truths are of a special kind: they depend only on logic, and not at all on the messy contingencies and accidents of haphazard phenomena on earth. Mathematics, Plato believed, paved the way to eternal truth, and in what other area could it be more appropriate than in understanding the cosmos? A divine creation, the universe must be perfect. Mathematics, with its irrefutable proofs, is a perfect form of knowledge. So mathematics is the key to understanding how the heavens work.

Plato's injunction that celestial architecture display an ideal mathematical form is all very well, but how is a philosopher to determine what particular kind of mathematics suits the job? The Pythagorean love of mathematical harmony, remember, began with the empirical discovery that musical harmonies embody simple arithmetical ratios. Plato regarded such practical concerns as beneath his dignity and therefore beneath the dignity of any true thinker of great thoughts. In the *Republic,* Plato conveys these criticisms through his teacher Socrates, a man who in truth was little concerned with what we might call scientific questions. When Plato has Socrates say something about the heavens, we can be sure we are hearing Plato's opinions. At any rate, Socrates has no time for bumblers who "measure the harmonies and sounds they hear against each other" and "look for numbers in these heard harmonies."[4]

Mathematical perfection, Plato-as-Socrates meant to say, is a self-evident ideal, not tied to this or that instance of it on earth. For Plato, the path to truth was reason and reason alone; empirical demonstrations were nice confirmations of mathematical principles in action, but they should not be taken as *justifications* of the ideal. In short, Plato put rational thinking first, with practical investigation a distant second.

Plato's admiration of mathematical and geometrical truths was the basis for his insistence that four of the regular geometric solids—the cube, the octahedron, the tetrahedron, and the icosahedron—were to be identified with the classical elements earth, air, fire, and water. The elements could then snug up against each other in a variety of ways, dictated by the laws of geometry, to form other substances.

Plato's bequest to astronomy was one we have already seen. According to the seventh-century CE commentator Simplicius, Plato asked philosophers to answer a fundamental question: "By the assumption of what uniform and orderly motions can the apparent motions of the planets be accounted for?"[5] It was unambiguous that "uniform and orderly motions" meant motion along circles at constant speed. Plato's direct contributions to astronomy were minimal, but he set in place the preoccupation of mathematical astronomers for many centuries: How can the apparently irregular paths of the sun, moon, and especially the planets be accounted for using circles, and only circles?

The ingenuity and even brilliance with which some geometers attacked this task is exemplified by the celestial model of Eudoxus, a student and follower of Plato. He put the earth at the center of the heavens and set about building a system of circles that would give the planets their observed motions across the sky. Take a transparent hollow sphere and mark a dot on its surface to represent a planet. Now set that sphere inside another sphere, on an axis allowing it to spin. Now put the second sphere inside a third, on an axis at a different angle, and put the third inside a fourth, on an axis at yet another angle.

Now suppose you are a keen observer on earth, at the center of these four nested spheres. Set the outer one spinning, the next spinning within that, the third within that, and the fourth within that—all at different speeds and at oblique angles. Eudoxus claimed that with four such spheres for each planet, and with three each for the sun and moon, he could do a decent job of reproducing the trajectories of the heavenly bodies.

I find it astonishing that Eudoxus, scratching marks on a slate or drawing figures in the sand, was able to determine from this model the paths that planets would make across the sky as seen from earth. For all that, alas, the model was only so-so. The paths of the five planets were variations on the same curves, and couldn't account for some of the differences that the actual planets' paths display. The biggest problem, though, was that Eudoxus had no explanation for why the brightness of the planets varied as they traversed the sky. Many early observers had jumped to the correct conclusion, which is that the distance between us and each planet is not constant. In

Eudoxus's scheme, though, each planet is fixed on a sphere set within other spheres, and all are centered on the earth. The planet's distance from us is unvarying.

When Aristotle came to write about astronomy, he confessed that the subject was not one of his strengths. In *On the Heavens,* he repeated Plato's insistence that celestial motions must be circular and described the complicated system of Eudoxus, which he then made still more complicated for reasons of his own that had to do with his thinking on the principles of motion, his insistence that a true void could not exist, and his desire to make the dodecahedron into the basis of the fifth element, quintessence, which filled space beyond the moon. We don't need to go into these intricacies. The upshot is that Aristotle created a planetary system with a total of fifty-six spheres (although some modifications could bring the number down to forty-nine or forty-seven) but, not being a mathematical sort, he provided no deep analysis of this model to show that it would work as desired.

To his credit, though, or perhaps in rueful recognition of the mess he had gotten himself into, Aristotle concluded his discussion of astronomical matters on a modest note: "If those who study this subject form an opinion contrary to what we have now stated, we must esteem both parties indeed, but follow the more accurate."[6] He left room, that is, for a better heavenly system to supplant the one he had described. And in due course, the Ptolemaic model, with its epicycles and equants, took over, and became part of the "Aristotelian" system of the Middle Ages. But Aristotle's concession that his might not be the last word on heavenly architecture was one that the later Aristotelians were decidedly reluctant to make.

To sum up, the philosophers of the pre-Christian era had a fervent interest in understanding the world and the universe around them, began to think of the physical world and the heavens as being governed by mathematical rules, and made serious attempts to account for the motion of the heavenly bodies. Their theories were mostly off the mark, by our modern perspective, and they spent a lot of time arguing about the nature of true knowledge and how to recognize it when they saw it—but this was an understandable preoccupation, given that they were the first to think seriously about how it is that we know anything at all.

What characterizes the ancient attempts to map out the heavens is a form of idealism. Because the heavens are the perfect creation of an infallible creator, they must be ruled by the most rigorous of intellectual systems, namely mathematics. And only the best mathematics would do. Circles were perfect, while other curves, although well-known to the ancients, were not. The underlying belief, moreover, is that human thought alone, powered by logic and reason, is what it takes to comprehend the heavens. Observation of the motion of heavenly bodies is important, to be sure, but pure reasoning is what allows us to make sense of what we see. The idea that we can understand the cosmos best through the application of pure thought is a deep-seated one.

A more appealing aspect of the ancient Greek philosophers is that they argued with each other, both contemporaneously, as they traded ideas back and forth and criticized each other's assumptions and methods, and also historically, in that later philosophers thought of themselves as building on the work of their predecessors, in some cases disagreeing greatly with their forebears but still pressing their attention on fundamental questions that earlier thinkers deemed important. They formed, in short, an intellectual tradition of great minds engaged in constructive debate and argument.

Which brings us back to the question that started this chapter. How did the lively Greek tradition of inquiry turn into the ossified dogmatism of the Middle Ages, and in particular into the orthodoxy of the Catholic Church?

4

The Holy Roman Empire Strikes Back

Rome was not destroyed in a day. The Roman Empire, beset by internal decay and barbarian outsiders, dwindled into insignificance by the fifth century CE, leaving western Europe a confused mess of warring kingdoms and city states. The Dark Ages overtook this part of the world, but elsewhere the lights stayed on. In the Arabic world, centered first on Damascus and later on Baghdad, and extending by stages into North Africa and then Spain, translations of some fraction of the intellectual works of ancient Greece were preserved, read, taught, and studied. From the eastern regions of their realm the Arabs acquired, among other things, a storehouse of astronomical observations from India and, famously, the "Arabic" numerals (actually a Hindu invention), including the previously unheard of concept of zero. In mathematics, the legacy of the Islamic world is still with us. We have "algebra," from *al-jabr*, an Arabic word meaning "reunion" that was part of the title of a book by the scholar al-Khwarizmi, from whose name we get "algorithm." (The reference to "reunion" comes about because al-Khwarizmi's book described systematic methods of combining equations in order to determine unknown quantities.)

The key figure in the slow emergence of western Europe from the Dark Ages is Averroës, who was born in 1126 in Cordoba, Spain. His westernized name derives from the last part of his Arabic name, ibn

Rushd. He was a learned man from a notable family, and by his thirties was working at the court of the caliph in Marrakesh, who governed what is now Spain and Morocco. The caliph, in his attempts to understand the constitution of the heavens according to the various philosophers of old, had trouble interpreting Aristotle. Averroës found himself tasked with writing an explainer, but what began as an assignment turned into a labor of love. He wrote a lengthy body of commentary on Aristotle's work, often amounting to line-by-line elucidation of the Philosopher's thinking. (Averroës referred to Aristotle as the Philosopher, as if there were no other contenders for that title.)

What attracted Averroës to Aristotelian thinking was the insistence that things happened for reasons, that nature conforms to laws written by God. This position put him at odds with another Islamic philosopher of the era, al-Ghazali, who a century earlier had attacked Aristotle and others in a book with the title *The Incoherence of the Philosophers*. Al-Ghazali argued for what is called occasionalism, which says that all phenomena on earth and in the heavens happen strictly because of God's will—which means that there are no rational laws of nature, since God can make anything happen at any time, according to His Whim. This belief became a fundamental tenet of Islam.

Averroës wrote a rebuttal of al-Ghazali with the acerbic title *The Incoherence of the Incoherence*. His arguments were heresy, according to the precepts that held sway in the eastern part of the Islamic world, but in Marrakesh and Spain, Averroës was at first safe from such charges. Toward the close of the twelfth century he was condemned and his books ordered burned, but a few years later he was back in favor with a new caliph in Marrakesh, and died there in 1198. As interest in Aristotelian thought grew, some of Aristotle's own works, which had been largely lost in their Greek and Latin versions, were revived from their Arabic translations.

In the rest of Europe, meanwhile, the power of Rome was gone but the Catholic Church remained. Many of the kingdoms that now controlled regions of western and central Europe had been founded by pagan invaders who became Christians and professed allegiance to the pope. The Lombards of northern Italy, for example, descended from the Langobardi, a crew of hairy heathens ("long beards") from Scandinavia, who swept south in the sixth century and, finding the climate and culture of Italy highly congenial, made themselves at home. Another Germanic people, the Franks, established themselves in what is now France at around the same time. Thus arose, in piecemeal fashion, the Holy Roman Empire (neither holy nor Roman nor an empire, as Voltaire famously put it), ruled by a Holy Roman emperor who acquired that title at the behest of the pope.

The Catholic Church was the glue that held the Holy Roman Empire together. What gave the church its authority was, in a literal way, the fear of God. The church could grant favors to the blessed and make life difficult—the afterlife especially—for the sinful. Fear of excommunication, fear of a life in hell, or of a long stay in purgatory before getting to heaven, lent genuine ferocity to the teachings and precepts of the church. Stalin (or perhaps Bismarck or Napoleon before him) was supposed to have asked how many divisions the pope could command. In the Europe of the Middle Ages, the answer was that he could command a great many. Kings and chieftains pledged allegiance to Rome and defended it against attacks and rebellions. The church's power took on a distinctly worldly aspect. Church officials from the pope on down accepted substantial donations and bequests in return for easing the transition of noblemen and wealthy traders into the arms of God. The church acquired land and money, and its splendor reinforced the impression of power.

As worldly as the church became—and by the year 1000 its avarice and corruption were notorious—its power rested ultimately on theology and, indirectly, on philosophy. For the threat of excommunication or condemnation to hell to instill genuine terror, there had to be stringent doctrines to distinguish the holy from the unblessed, the saint from the sinner. The philosophers who, starting in the twelfth

century, provided the intellectual foundation for Catholic orthodoxy became known as the Scholastics. They held Aristotle in great
esteem and particularly admired his attention to the intricacies of
logical deduction and syllogistic reasoning. In the words of Bertrand
Russell, "The general temper of the scholastics is minute and disputatious rather than mystical."[1] In other words, they were less about
grand principles and cosmic questions than about rules and regula-
tions and how to correctly apply them.

The most notable of the Scholastics was Thomas Aquinas (1225
or 1226 to 1274), whose four-volume *Summa contra gentiles* was
intended as a rational argument for Christian belief. Aquinas was
greatly influenced by Averroës, and like him referred to Aristotle as
the Philosopher; Averroës himself, in Aquinas's writings, becomes
the Commentator. These names hint at something fundamental:
argument from authority. Catholic orthodoxy was to a large extent a
matter of insisting on certain beliefs because the Philosopher or the
Commentator had insisted on them. Aquinas's philosophy, dubbed
Thomism, combined ideas from Aristotle along with new interpre-
tations of Plato (Neoplatonism). Pronouncements on motion come
from Aristotle and the cosmology is Aristotle's retelling of Eudoxus.

Shortly before Galileo's time, in 1517, Martin Luther nailed his
ninety-five theses to the church door in Wittenberg, marking the
beginning of Protestantism. The response of the Catholic Church
was to double down on orthodoxy and the prosecution of heresy
(the Inquisition had been founded by Pope Gregory IX in the early
thirteenth century for that purpose and, after Luther, found much
new work). At the Council of Trent, a meeting of church leaders
that occupied twenty-five sessions over the years 1545 to 1563, church
leaders hammered out a long list of arguments showing why the
Protestants were wrong and why Catholic doctrine was correct. The
resulting Tridentine Creed standardized Catholic beliefs and led to a
revised catechism and a Tridentine Mass that remained cornerstones
of Catholicism for centuries afterward. The Council of Trent also

ordered the creation of the *Index Librorum Prohibitorum*, the index of prohibited books, by which heretical writings were censored and suppressed. If the church was seen to be riven by dissent, split into antagonistic factions, then its position as the arbiter of God's truth, the guide to the one true path to heaven, would be undermined. So it ruthlessly stamped out heresy.

We can now understand better how the church in Galileo's time had such a strong hold not only on the mundane activities of life but on the life of the mind, too. It was not simply that Aristotelian thinking reigned supreme, but that reverence to long-standing orthodoxy was deeply entrenched. The very idea of learning consisted largely in understanding what the Philosopher said, what the Commentator thought about it, and how Saint Thomas Aquinas (beatified in 1323) tied it all together with Christian doctrine.

On the other hand, having bought Aristotelianism wholesale, so to speak, the pope and his advisers could afford to exercise a degree of prosecutorial discretion in deciding who to go after and who to leave alone. Galileo's work on the motion of cannonballs and falling stones contradicted Aristotelian theories of motion but did not get him into any great trouble. It was only when he overtly embraced Copernicanism—with its essential elements of a stationary, central sun and a moving earth—that the church began to think he was entering heretical territory. Heliocentrism ran counter to the Aristotelian picture of the heavens, but it was also a case where the Bible had something to say. The crucial evidence came from Joshua, chapter 10, wherein the Lord assists Joshua and his men in smiting the forces of the five kings of the Amorites. The Lord starts things off by sending down great stones and hail on Joshua's enemies:

[12] *Then spake Joshua to the Lord in the day when the Lord delivered up the Amorites before the children of Israel, and he said in the sight of Israel, Sun, stand thou still upon Gibeon; and thou, Moon, in the valley of Ajalon.*

[13] *And the sun stood still, and the moon stayed, until the people had avenged themselves upon their enemies. Is this not written in the book of Jasher? So the sun stood still in the midst of heaven, and hasted not to go down about a whole day.*

If God made the sun stand in the sky, then it clearly must have been moving beforehand and would resume moving afterward. For decades after the 1543 publication of *De revolutionibus,* the church had managed to go along with Copernicus on the presumption that sun-centered geometry and a moving earth were to be regarded purely as mathematical devices and certainly not understood as natural truth. In doing so, the church leaned heavily on the preface to *De revolutionibus,* which unequivocally stated the book's hypothetical nature.

But this convenient subterfuge was dispelled by Johannes Kepler in his *Astronomia nova* of 1609. He had found out that the preface to *De revolutionibus* had not been written by Copernicus after all, but came from Andreas Osiander, a German theologian who had helped guide the book to publication. The preface was Osiander's way of safeguarding Copernicus's system from objections he knew it would encounter. Having blown the whistle on Osiander, however, Kepler forcefully argued that Copernicus meant his model as the literal truth. Against this new provocation the church could no longer sit idly by. In 1616, it placed *De revolutionibus,* which had circulated freely for well over half a century, on the *Index Librorum Prohibitorum,* and in 1620 it released a sanitized version, with ten emendations, that was acceptable to the church. The subject of heliocentrism was strictly off-limits to the obedient: they should not read about it, talk about it, and especially argue for it. Galileo's advocacy of Copernicanism, veiled though it had generally been, came under ever greater suspicion.

But the power of the church was not as great as is commonly supposed. Owen Gingerich, through his lifelong effort to track down and inspect as many copies of *De revolutionibus* as he could find, discovered that the papal ban on the book had limited consequences. He found that Galileo had dutifully censored his own copy and that about two-thirds of the copies in Italy showed the required corrections. But in other countries, even in strongly Catholic France, hardly anyone bothered to make the corrections. In Spain the local version of the *Index* continued to allow *De revolutionibus*.[2] Readers in Protestant countries paid no heed to papal edicts.

In Italy, though, Galileo was obliged to tread carefully. In private, he suggested that the verses from Joshua were entirely consistent with the notion of a stationary sun and a rotating earth: if God had chosen to make the earth stop spinning for a day, the effect would have been exactly the same as if he had temporarily paused the motion of the sun around a stationary earth. Here, though, the longstanding belief that the earth could not possibly be moving, because mighty gales would blow constantly and buildings could not stay upright, was entrenched in the minds of conventional philosophers.

Galileo also argued (taking a cue from Saint Augustine) that while the Bible was surely the supreme authority on matters of morality, if there was a conflict between Scripture and a question of empirical truth, then it was permissible to treat Bible stories as metaphorical. Aquinas, however, had rejected this position, saying that the words of the Bible could never be contrary to reason: the meaning might sometimes be obscure but it could always be found by sufficiently astute philosophers. The Council of Trent had forcefully declared that interpreting the Scriptures was a job for experts only. What the Bible said would in all cases be what the church in its wisdom said that it said.

The clash between Galileo and the church thus begins to seem inevitable. Simplistically, it was a clash of the new with the old, of reason against faith, of open-minded investigation upsetting the established orthodoxy. Still, four centuries later, almost every account of the story provides a different coloration of the basic black-and-white picture and a different emphasis on the roles and motives of the participants.

Stillman Drake, a historian of science who devoted much of his academic life to the study of Galileo, argued that Galileo and the church might well have contrived to get along, a little awkwardly but without open conflict, had it not been for the hard line taken by certain theological philosophers. In this version of the story, Galileo—"a prudent man, not given to forming conclusions without having weighed the evidence, well aware of social customs, and disinclined to quarrel with highly placed persons in church or state"[3]— was a devout Catholic who was slow to embrace heliocentrism and who feared that the church was making a grave error by tying its flag to any particular hypothesis about the architecture of the heavens, whether Aristotelian, Ptolemaic, Copernican, or anything else. Though he was eager to set aside the erroneous tenets of Aristotelian natural philosophy, Galileo had no desire to establish a competing grand philosophy of his own. He foresaw that science would proceed bit by bit, one observation, one theory at a time, and would in that way incrementally build up a picture of the natural world. No seeker of the truth should construct a cosmology on flimsy and incomplete foundations, because new discoveries would necessitate constant tinkering and adjustment. That was equally true, Galileo believed, for a scientist, a philosopher, or a churchman. Tacitly, he hinted at a deal: when it came to natural phenomena, he would argue only for those ideas for which he had good evidence and reason, if the church would do the same. According to Drake, there were senior figures in the church who thought along the same lines. Indeed, Galileo's early discoveries with the telescope, recounted in *The Starry Messenger*, had been generally celebrated by the church. Local boy makes good: here was an Italian, a Catholic explorer, a friend to numerous priests and cardinals, a man whose accomplishments led to an audience with Pope Paul V, enlarging our vision of the universe, adding thereby to the glory of God.

There were difficulties, to be sure. Through his telescope, Galileo had clearly seen bumps and irregularities at the edge of the moon's disk. It was not, evidently, a perfect sphere. Some theologians responded that in fact the moon was a perfect sphere, but one made of a transparent substance containing irregular opaque chunks. By and large, though, the church did not object strenuously to Galileo's findings from his lunar observations. Cardinal Robert Bellarmine, a learned and influential church figure who at first represented a moderate viewpoint that was helpful to Galileo, raised no objection to their publication. The Bible, after all, said nothing directly about the sphericity of the moon; this was an issue that didn't raise too many hackles.

Far more troublesome was the heliocentric system. Galileo went to Rome in 1615, hoping to argue his view that the church could afford to exercise some flexibility on the question, but Pope Paul V was more concerned with the church's worldly standing than its position on the heavens. He was surrounded by advisers who cautioned him that tolerating heretical views even on arcane matters would be taken by the enemies of the church as an acknowledgment of weakness.

Thanks to the intervention of Bellarmine, who respected Galileo, the Inquisition was for the moment held at bay. Instead, Bellarmine persuaded the pope to ask a committee of eleven theologians to examine two questions—Is the sun central and stationary? Is the earth displaced from the center and in motion?—and report back. In early 1616, they did so. The first proposition, that the sun does not move, was found to be "foolish and absurd in philosophy," and "formally heretical inasmuch as it contradicts the express opinion of Holy Scriptures in many places." The second proposition, that the earth moves, received "the same censure in Philosophy, and with regard to Theological verity . . . is at least erroneous in the faith."[4]

Drake says that Galileo was surprised and disappointed by this decision because he had expected the theologians to conclude that the church had no business passing judgment on matters of astronomical investigation. But this seems naive (whether on Drake's part or Galileo's I am not sure). The church and its theologians and philosophers had a long history of making pronouncements on the nature of the universe. The science historian John Heilbron, on the other hand, emphasizes that while theologians found it heretical to say that the sun is stationary, they concluded it was merely "erroneous in faith" to say that the earth moves, because the Bible has nothing to say that bears directly on that point. For that reason, says Heilbron, it's incorrect to say that the church deemed Copernicanism heretical.[5] But this is stretching a point: in the Copernican model, taken at face value, the sun is stationary and the earth moves. You can't have one without the other.

In the late nineteenth century, Pierre Duhem, a French physicist turned historian and philosopher of science and an avowed Thomist, offered a different perspective. Galileo, Duhem said, went too far in extolling the truth of the heliocentric system at a time when the evidence for it was decidedly shaky. Bellarmine, by contrast, wrote a letter that Duhem said was "full of wisdom and prudence," in which he stuck to the old line that it was acceptable to treat Copernicanism as a mathematical hypothesis but that insisting on its reality was too much.[6] Duhem did not doubt that heliocentrism was right, only that Galileo jumped the gun. He argued that when the church admonished Galileo to be less vehement in his advocacy of the new astronomy, it was not being reactionary but properly cautious in its evaluation of a contentious proposal. This idea still surfaces from time to time, for example in the writings of John Polkinghorne, a Cambridge theoretical physicist who gave up research to become (in 1982) an Anglican priest,[7] and in *Science and Religion,* a 2004 study by historian Richard G. Olson.

Was the church acting more *scientifically* than the reckless Galileo? The assertion doesn't hold up. It was Galileo who had been cautious about Copernicanism until he had real evidence, most strongly from the phases of Venus, and by the early seventeenth century, the church's opposition derived explicitly from the contradiction with Scripture.

There's one more strand to the story that deserves brief attention. Even allowing that Galileo was right and the church wrong, some commentators have suggested that Galileo was arrogant, conniving, egotistical, sneaky—that the church strove to be reasonable but Galileo repeatedly and willfully tested the limits of whatever freedom he was granted. Heilbron says that when Galileo debated controversial issues by means of dialogues, he deliberately made the speakers of views opposed to his look foolish and obtuse. It was not the Inquisition that deployed *sarcasm* but Galileo. Sometimes he wrote pamphlets anonymously, in the Paduan dialect of Italian, to make his critics look like rubes and yokels.

Behind these assertions lurks a counterfactual history in which Galileo, if only he had been more measured, more circumspect, more respectful in his deeds and writings, would have got his ideas across without angering the church to the point of vengeance. It's impossible to say, of course. But ultimately, as Bellarmine himself recognized, the dispute came down to questions on which either Galileo or the church was right: Is the sun stationary? Does the earth move? There was no middle ground.

As the dispute simmered uneasily, Galileo wrote two new books that stand as milestones in the history of science. *Il Saggiatore* (*The Assayer,* 1623) offered a defense of what we might loosely call the scientific method, favoring reason based on observation and experiment over mindless recitation of the "truth" found in old books of philosophy. *Dialogo sopra i due massimi sistemi del mondo, tolemaico e copernicano* (*Dialogue Concerning the Two Chief World Systems, Ptolemaic and Copernican,* 1632) makes a clear case for the superiority of the Copernican system over the Ptolemaic on its agreement with a wide range of observationally established facts, regardless of ancient prescriptions of how the sun and the earth and the planets "ought" to move.

Galileo had no trouble getting *The Assayer* published, and its slashing style greatly amused Pope Urban VIII, who, before attaining the papacy in 1623, was Cardinal Maffeo Barberini, a Florentine nobleman and an old friend and admirer of Galileo. Publishing the *Dialogue*, however, demanded care. Galileo met with his friend the pope and, in Heilbron's reading of the event (no records survive), convinced Urban that a thorough, reasoned presentation of the case for Copernicanism—strictly, of course, *as a hypothesis*—would be of value to the church, in that it would allow theologians and philosophers to marshal their most persuasive counterarguments. The church could then be seen to be engaging in serious intellectual debate rather than acting out of mere mulishness.

This strategy proved acceptable to the Inquisition, but it insisted on certain amendments. One was that counterarguments to Copernicanism should be plainly presented. On this point, Heilbron's assertion that Galileo was too brash for his own good gains some credence. Galileo found rhetorical ways to make sure that, no matter how evenhanded the dialogues might appear to be, the pro-Copernican side would always come out looking better. The *Dialogue* was a lengthy back-and-forth between three characters. There was the ingenious and resourceful Salviati, the stand-in for Galileo himself, who made the case for Copernicanism and cleverly rebutted counterarguments. There was Sagredo, exemplifying the man of reason who, initially taking no side, came around to the Copernican viewpoint after duly considering Salviati's spirited reasoning. And there was Simplicio, representing the old guard, who dutifully parroted Aristotelian dogma only to be knocked down smartly by Salviati and made to look dim in the process. Though cast as a debate, the *Dialogue* left no attentive reader in the dark about who was right and who was wrong.

A second requirement was that the *Dialogue* must include an argument from Urban VIII himself that God was omnipotent and that no matter how convincing the Copernican system might seem, he would always have some divine Wiggle Room to run the heavens differently. Galileo was supposed to put this statement in his own words, says Heilbron, but he couldn't bring himself to do it, and so put it into the mouth of Simplicio. The book went to the presses early in 1632, but Urban instantly saw that Galileo was mocking him. In July he ordered the book suppressed. A panel was established to inquire further into the book's contents, and in the course of its investigation it unearthed documents concerning the nature of the warning that had been given to Galileo in 1616, after the committee of theologians had rendered their verdict. In short, Galileo had at that time received a private message from Bellarmine instructing Galileo that he could no longer avow or defend Copernicanism, but leaving him with the impression that he could still teach the subject, as long as it was presented as hypothesis. In the Vatican records, however, lay an official notice instructing Galileo that he could not avow, defend, or teach the heretical system. But this note was not properly signed; Bellarmine had refused to put his name to it, because it went beyond what he believed the Inquisition had decided.

Galileo was summoned before the Inquisition in 1633. By this time Bellarmine was dead, but Galileo was able to produce a letter signed by Bellarmine that said nothing about him being forbidden from teaching Copernicanism. Against this was the unsigned note in the Vatican archive that forbade him from doing so. Galileo was now sixty-nine years old and in failing health. Under questioning, he claimed to have trouble recalling what he had been told so many years ago. It was these inconsistencies, rather than any deep scrutiny of heliocentrism, that exercised the Inquisition—bolstering Bertrand Russell's perception of Catholic philosophy being "minute and disputatious" rather than interested in grand themes.

The Inquisition clearly doubted Galileo's honesty, but, given his age and fame and his good relations over the years with a number of influential churchmen, his punishment was not as harsh as it might have been. Galileo was asked to confess his vanity and to acknowledge his heresy, and on June 21, in what must have been a hard moment for him indeed, Galileo swore that "with a sincere heart and unfeigned faith I abjure, curse, and detest the [Copernican] errors and heresies."[8] He was then delivered into house arrest in the relatively benign surroundings of Arcetri, not far from Florence.

Supported by friends, Galileo was persuaded, while at Arcetri, to compose his *Dialogues Concerning Two New Sciences,* which included his final thoughts on the nature of motion and his explanation of the path of a cannonball. His health steadily worsened, and he died in January 1642, at the age of seventy-seven.

This condensed account of Galileo's tribulations should make it clear that there is no simple explanation for how they came about. At the bottom of it all were indeed the two questions Cardinal Bellarmine asked in 1616: Does the earth move and is the sun stationary? But overlaid on that were issues of friendships and rivalries, broken friendships and personal vendettas, factions within the church. . . . Galileo was caught up in a power struggle that played out partly on account of his own views and partly at his expense, as other figures in the church sought to advance their positions.

What mattered, in the end, is who had the final say. For the church, it was the language of the Bible that held ultimate authority, although in reality it was the language of the Bible as interpreted by the church's preferred scholars. Intellectual exploration of the cosmos was not in itself a rebellious activity—but the goal, in the eyes of the church, was to find further reason to support accepted beliefs, not subvert them. This was the essence of Thomist thinking: the church is right, and the purpose of philosophical investigation was to arrive at that conclusion in new and more convincing ways.

Galileo, though, had in mind a different kind of authority, larger than any personal or institutional opinion. *The Assayer* includes a famous statement:

Philosophy is written in this grand book—I mean the universe—which stands continually open to our gaze. It is written in the language of mathematics, and its characters are triangles, circles, and other geometrical figures, without which it is humanly impossible to understand a single word of it.[9]

The church had the Bible and a dogmatized form of Aristotelianism. Galileo had mathematics. Exaltation of the power of math to disclose cosmic truths was one of Plato's cherished principles, of course, so Galileo's statement has been interpreted to mean that he was rejecting Aristotle only to put Plato back on top of the heap.

But this is a gross misrepresentation. Galileo's attitude toward the value and utility of mathematics was wholly different from Plato's, and to understand how modern science emerged and flourished, we need to understand that Galileo moved beyond Plato and Aristotle in equal measure.

5

How Science Uses Mathematics

Plato prized mathematics for its pure logic and rigor. It begins with clearly stated assumptions or axioms and establishes theorems of irreproachable validity. A mathematical truth is intrinsically and eternally true, and depends not at all on the haphazard and chaotic phenomena of the natural world. That's all well and good, but where Plato went next was a leap—a leap of faith, if you like. If the universe itself is perfect and eternal, as it surely must be, then the only truths humans can know that are of equal grandeur are mathematical truths. Therefore, Plato concluded, only mathematics is fit to describe the cosmos, and only in mathematics can we hope to find truths that will express the architecture of the heavens.

Very grand, but what next? Mathematics, even in Plato's day, appeared to be a subject unconstrained by worldly limitations, containing multitudes of true things. So there was another leap. The perfection of the cosmos required that it use only the most perfect elements of mathematics: circles and the regular Platonic solids. Any other forms of geometry would sully the celestial realm.

It is very obvious today that this is not science, or even an ancestor of science, but a form of philosophical idealism. It relies, moreover, on human judgments of what elements of math are the most worthy of our respect, and takes the Creator's agreement for granted. Not to mention that the whole argument is fundamentally circular. We require the heavens to be perfect, we deem certain parts of math more perfect than others, and lo—we have the ingredients for cosmology. The cosmological conceits of Platonism amount to wishful thinking. We want things to be so, so they must be so.

Galileo determined that the path of a flying cannonball was a parabola. In doing so, he broke with the purest form of Platonism in two ways. First, he made use of a curve that ancients knew but regarded as imperfect. Second, he deduced the parabolic form by combining what he had observed of the behavior of falling objects with some simple reasoning. On this second point he set aside the Platonic stricture that mathematics alone should tell us how the physical world behaved.

Of course, Galileo was dealing with a terrestrial phenomenon, not a heavenly one, so perhaps he felt less constrained by traditional philosophical principles. Still, it's an intriguing question why Galileo, after so many centuries, was the first to break through the old dogma and arrive at the straightforward—to the modern eye—position that we can understand the physical world by combining experimental and observational data with a judicious use of mathematics applicable to the case at hand, not because of some overarching but vague belief in the intrinsic power of math by itself to guide a thinker to the right answer.

An equally intriguing answer to this question is that Galileo's thinking may have taken cues from his father's expertise in music—in particular, the theory of musical harmony. Vincenzo Galilei had been a student of Gioseffo Zarlino, whose 1558 book *Le istitutioni harmoniche* (which can be translated as "The Principles of Harmony") was the most sophisticated treatise of its time. Zarlino professed Pythagorean musical theory with some modern amendments. The original Pythagorean system of harmony had three basic musical intervals, each defined by the numerical ratio between lengths of string, say, that would produce those notes. There was the octave, a 2:1 ratio; the perfect fifth, 3:2; and the perfect fourth, 4:3.

The undoubtedly apocryphal story is that Pythagoras first heard these intervals when he was passing by a smithy in which the blacksmiths were banging their hammers on anvils and producing pleasing harmonies. I say apocryphal because it stretches credulity to imagine that a company of ancient blacksmiths would have accidentally acquired a set of anvils tuned to musical tones, still less that they would have spent hours whittling their anvils to the correct sizes to produce these perfect intervals. (If there really were any such blacksmiths, they would deserve to be called the first Pythagoreans.)

At any rate, Pythagorean harmony yielded a variety of implications, some useful, some decidedly awkward. A combination of a fifth and a fourth produces an octave: $3:2 \times 4:3 = 12:6 = 2:1$. And the difference between a fourth and fifth is $3:2$ divided by $4:3$, which is $9:8$; this is a whole tone interval. However, an interval of two whole tones is $9:8 \times 9:8 = 81:64$, which is almost but not quite the interval now known as the major third, $5:4 = 80:64$. That's a small difference but one that trained musicians and sensitive listeners could hear. By Zarlino's time the major third and other intervals had been added to the basic system of harmony. There was the minor third ($6:5$) and the major and minor sixth ($5:3$ and $8:5$). Zarlino wrote the last one, however, as $(2 \times 4):5$, because he felt it was wrong to include the number 8.

In his enlightening book *Galileo's Muse,* Mark Peterson of Mount Holyoke College makes the point that these small discrepancies didn't create insuperable problems for musicians until the blossoming of polyphonic music around the fifteenth century. When two or more notes are sounded together, they have to harmonize in a way that pleases the ear. Conformity to simple numerical ratios didn't always do the job. While Zarlino promulgated a modified Pythagorean system, Vincenzo Galilei, who was a professional musician as well as a student of theory, came to the conclusion that what musicians did in practice was not what Zarlino said they did, or should do. Vincenzo argued that musicians, by ear, adopted a scale in which each of the notes of an octave was separated from the next by the same interval—which meant that none of the intervals were perfect according to Pythagorean accounting.

Vincenzo discovered that an ancient writer, Aristoxenus, had come to the same conclusion, only to be roundly criticized by Socrates. Performers who tuned their instruments by practice rather than according to the principles of philosophy were people who "prefer their ears to their intelligence."[1] Recall that in chapter 3 we heard Socrates (according to Plato, anyway) being scathing about those who "look for numbers in these heard harmonies."

According to Peterson, Plato criticized Pythagoras for being too eager to understand the empirical practices of working musicians rather than focusing strictly on the elementary numerical ratios that, as far as Plato was concerned, were the only way to truly understand harmony as the Creator intended it. In short, it was Plato who made a fetish of reducing musical harmonies to simple numbers, in the process doing his best to erase the work of earlier analysts who understood the questions better. (To be fair, the history is murky: Daniel Heller-Roazen, in *The Fifth Hammer: Pythagoras and the Disharmony of the World,* leans to the view that the Pythagoreans were well aware of the difficulties in fitting actual musical harmonies to their excessively simple scheme, but did their best to hide the problems.)

The relevance of this musical digression is that Galileo, around the age of twenty, is thought to have participated with his father in experiments on vibrating strings that demonstrated the inadequacy of Zarlino's teachings on harmony. Early in his life, therefore, he became aware of an incontrovertible case in which the desire of Plato and others to make reality conform to idealized mathematics simply didn't work. He learned that if experiments, as well as the real-life habits of musicians, were at odds with the strictures of philosophy, then one had direct and trustworthy evidence that philosophy was wrong.

We see this lesson repeated in Galileo's recognition of the parabola. Daniel Santbech, as we saw in chapter 1, was acutely aware that cannonballs did not appear to follow the path prescribed by philosophy: a straight line rising at some angle, followed by a vertical drop. His conclusion was that since philosophy was incontrovertible, then our senses must deceive us in some way. Vincenzo Galilei and his son, when it came to musical harmony, came to what we now see as the obvious conclusion: it was philosophy that must be wrong. So Galileo, when he came to think about the path of a cannonball, and primed by his many earlier experiments on falling objects, had no qualms about saying that philosophy was wrong in this case, too.

How, then, did Galileo think about mathematics? He certainly understood and admired its rigor. Peterson offers a remark from the *Dialogue Concerning the Two Chief World Systems* when Salviati (Galileo's stand-in, remember) says of mathematics that it is a kind of knowledge "in which the Divine intellect indeed knows infinitely more propositions, since it knows all. But with regard to those few which the human intellect does understand, I believe that its knowledge equals the Divine in objective certainty."[2] Galileo is saying, that is, that we poor humans cannot know as much math as the Creator, but where we do know it, our certainty is as great as his.

This seems like a Platonist thing to say, in that Galileo is extolling math for the absolute truth it can convey. But that is only the starting point for full-on Platonism, which goes on to assert that only by contemplating mathematics, and mathematics alone, can we hope to arrive at comparably certain truths about the nature of the universe. On this aspect of Platonism, Galileo clearly didn't agree.

In the *Dialogues Concerning Two New Sciences,* Sagredo at one point remarks, "Logic, it appears to me, teaches us how to test the conclusiveness of any argument or demonstration already discovered and completed; but I do not believe that it teaches us to discover correct arguments and demonstrations."[3] Sagredo, having listened carefully to Salviati's arguments, is coming round to the latter's point of view: logic, which is to say mathematics, can test arguments but cannot by itself create new thinking about the physical world. That's where observation and experimentation are indispensable.

Was that Galileo's final word on the matter? Perhaps not. According to biographer John Heilbron, Galileo as an old man acknowledged the importance of mathematical reasoning in distinguishing logical from illogical propositions but went on to say that "this use of mathematics as a sieve for removing crude ideas does not meet my ambitions. Mathematicians should not merely enforce right thinking among philosophers, they should initiate and guide it."[4]

Now, it seems, Galileo is saying mathematics *can* lead us to the truth, as Plato insisted. But there's a problem with this remark: *Galileo didn't say it! Heilbron made it up!* Heilbron imagined a dialog between old Galileo and young Galileo, in which the former is instructing the latter on his mature opinions on the method of science. Just as Socrates comes to us through the decidedly biased mouthpiece of Plato, so in this instance Galileo is speaking what Heilbron believes he should have said, close to the end of his life. Heilbron has, to my mind, a strange animus against Galileo, and constantly depicts his arguments with authority in the worst possible light. In this case, he goes even further, making Galileo utter Heilbron's opinion of what his philosophy of science ought to have been, and making him in the process a closet Platonist. This is a strange way for a historian of science to carry on.

Let us return to the famous comment that Galileo, almost sixty years old, really did make in *The Assayer*: "Philosophy is written in this grand book—I mean the universe—which stands continually open to our gaze. It is written in the language of mathematics, and its characters are triangles, circles, and other geometrical figures, without which it is humanly impossible to understand a single word of it."

This paean to the power and importance of mathematics has been adduced as proving that Galileo "belongs to the Platonist camp."[5] The Russian-born French philosopher of science Alexandre Koyré went further. He claimed that Galileo had never done his famous experiments on falling stones and balls rolling down inclined planes and insisted instead that they were "thought experiments" conducted in his mind only to bolster conclusions that he had already reached by mathematical reasoning alone. (It's worth noting that "thought experiments" as an element of scientific reasoning were unheard of until Einstein came on the scene and analyzed minutely what would be seen by an observer on a train traveling close to the speed of light, for example; it's absurdly ahistorical to suggest that Galileo indulged in the same kind of reasoning three centuries earlier.) Koyré then went on to assert that Galileo was an out-and-out Platonist, reacting against the Aristotelian orthodoxy of his day and further implies, among other things, that Galileo was obliged to choose one or the other of the ancient dogmas and was incapable of forging his own path. Through study of Galileo's notes, however, Stillman Drake made a convincing case not merely that Galileo had conducted experiments but that he had done so with great care and attention to sources of error and confusion.[6]

I see no reason at all to agree with philosophers and historians who, for reasons of their own, want to turn Galileo retroactively into a Platonist. But what, then, did Galileo mean with his statement about the "language of mathematics"? The reference to "triangles, circles, and other geometrical figures" is somewhat misleading to the modern reader. In Galileo's time, geometry was the most esteemed form of mathematics, and Galileo's theorems about motion were written as geometric propositions that now seem needlessly laborious. But that was how it was. Forms of mathematics more suitable for problems in physics remained in the future.

My interpretation of Galileo's meaning comes down to two points. First, he is certainly saying that if we want to describe the universe in a scientific way, we need to describe it quantitatively, precisely, and logically—that is to say, in the language of mathematics. But there's a second phrase that gets less attention. In his first sentence he emphasizes that the universe is "continually open to our gaze." He means, literally, that we are able to observe it and that this must be the foundation for any mathematical description of it. Plato insisted that it was enough to *think about* the universe, and that a philosopher would in due course be able to deduce thereby its true mathematical form. Galileo said that we also need to *look at* the universe, and from our observations infer our way to the most appropriate mathematical description.

This is a crucial distinction. It's why Galileo was not a Platonist in the idealistic sense but a thinker who prized empirical information as the starting point for any true scientific investigation. His insistence on careful and minute examination of the evidence around him would have pleased Aristotle, who loved to observe and form hypotheses but who, in the biological arena that he favored, had no means to devise crucial tests. In this sense, Galileo took elements from both Aristotle and Plato and artfully combined them to create what we recognize as science. His style of investigation, moreover, insists that mathematics must be the servant, not the master.

PART II

CLASSICAL SCIENCE REIGNS SUPREME

Tthere was no such thing as the Scientific Revolution, and this is a book about it." With this cute remark Steven Shapin, a historian of science, begins his 1996 book *The Scientific Revolution,* the purpose of which is to say that the emergence of modern science during the Renaissance was not revolution but evolution. Historians and sociologists of science, as far as I can tell, have an ideological aversion both to great man narratives, in which major changes are largely due to a few people, and to depictions of change as revolutionary, because, if you look hard enough, you can always find antecedents and premonitions of disruption. There's some truth here: science is a collective enterprise, and new ideas do not emerge like rabbits from hats. But Shapin's critique goes too far. Certainly, if you examine the intellectual life of the Renaissance from day to day, you will not find a specific hour when prior beliefs were overthrown and new ones installed in their place. Even revolutions take some time. The fact is, though, as I hope the previous chapters have made clear, Galileo played a huge role in upending old philosophy that clung to orthodoxy and authority and opened up a new world in which observation and reason were of the essence.

If you want a thorough refutation of Shapin's thesis, I strongly recommend a recent book by David Wootton, *The Invention of Science*. That said, let us turn to the consequences of the Galilean revolution.

6

Mastery of Motion

The ancient Greeks could not understand motion. Not the actuality of it, of course—the world was full of things that moved, and even philosophers were capable of perambulating from one comfortable thinking spot to another—but the *idea* of motion. It seemed impossible to them, in a way that was exemplified by the famous paradox attributed to Zeno of Elea. In the version I first heard, a frog wants to jump into a pond a short distance away. First it has to jump half the distance, then half the distance that remains, then half of that remainder. . . . And so, said Zeno, the frog must make an infinite number of jumps, which means it can never reach the pond.

Galileo made great progress in understanding motion but was hampered by his lack of mathematical tools. The third day of the *Dialogues Concerning Two New Sciences* begins with a list of some theorems about motion at constant speed that strike the modern reader as absurdly long-winded and repetitive. One theorem states that if a body moving at a certain speed travels a known distance in a known time, then it will travel twice that distance in twice the time. Another says that a body moving twice as fast as another will travel twice the distance in the same time. Another says that if one body travels a given distance in half the time that another takes, it must be moving twice as fast. . . .

To the modern reader, Galileo is saying the same thing over and over: a constant speed is simply distance traveled divided by time taken; or, to invert the relationship, distance is speed multiplied by time. In the language of algebra: $v = d/t$; $d = vt$. Galileo's theorems are straightforward applications of these basic definitions.

But Galileo didn't have algebra, or rather, he avoided it. Algebra, courtesy of the Arabic mathematicians of some centuries earlier, was known in Galileo's time but was not respected. Algebra was the mathematics of shopkeepers and quartermasters, useful for working out how to price items or how much bread to buy for a battalion of soldiers, but it lacked the purity and philosophical rigor of classical geometry, the language in which eternal mathematical truths were traditionally expressed. To prove his theorems about uniform motion, Galileo devised a different geometrical construction for each one, so that the relationship between them was not clearly evident.

Despite this handicap, Galileo pushed on into the far more tricky case of a body moving with constant acceleration, as in his much-studied examples of objects falling to earth. They start out slowly and get faster. That's what acceleration means—but there was an issue to decide. Does this constant acceleration mean that the body picks up a fixed increment of speed in equal time steps, or a fixed increment of speed in equal distances fallen? Guided by his careful experiments as well as mathematical reasoning, Galileo concludes that the first definition is the correct one.

There ensues a discussion, in dialogue form, of how constant acceleration can be analyzed and deductions made about the speed and distance traveled by a body moving in this way. Galileo has Sagredo (the mouthpiece of traditional philosophy) articulate a version of Zeno's paradox: working backward, Sagredo says that a body moving at some speed must have acquired half of that speed in half the elapsed time, a quarter of that speed in a quarter of the time, and so on down to limitlessly small increments of speed and time. Sagredo then turns the argument around and asks how a body initially at rest, with no speed at all, can ever move off its mark. It seemingly needs to take an infinite number of steps to get moving at all.

Salviati, Galileo's alter ego, replies that his experiments on falling bodies have shown him, as a matter of empirical fact, that an object dropped from a height as tiny as you can imagine will indeed acquire a correspondingly tiny speed. That's all very well, Sagredo objects, but it doesn't answer the philosophical question of how it happens.

Now Salviati makes the crucial observation: the falling body does not acquire a certain speed and stay at that speed for a finite moment. Rather, "it merely passes each point without delaying more than an instant: and since each time-interval however small may be divided into an infinite number of instants, these will always be sufficient to correspond to the infinite degrees of diminished velocity."[1]

The wording is opaque, but what Salviati, I mean Galileo, is saying is that a succession of infinitely small increments of speed occurring in a series of infinitely tiny intervals of time can add up to a finite speed. In the same way, what Zeno failed to imagine is that the infinitesimally tiny jumps that the frog makes on its way to the pond occupy infinitesimally tiny amounts of time—they are not discrete events, each requiring some perceptible time in which to occur. Galileo could see qualitatively how to get around paradoxes of this sort and, impressively, he used his geometrical methods to prove some theorems concerning motion with constant acceleration. Even so, constant acceleration is a special case of motion in general. A body may slow down, speed up, slow down again in all kinds of ways. Its acceleration, like its velocity, may vary from moment to moment. Galileo could see no way of dealing with such complexity. That task fell to Isaac Newton, who was born in 1642, the year Galileo died.

Newton was an extraordinarily strange character. Lacking, it appeared, any need for human companionship, he worked obsessively in isolation, and having solved a problem to his satisfaction, put the answer aside and moved to another obsession. But word of his genius trickled out despite Newton's apparent disdain or fear of making his discoveries known.

Newton's father was a farmer from Woolsthorpe, Lincolnshire, sixty or so miles north of Cambridge, who could not sign his own name and who died before his son was born. His mother had connections, in particular a brother who had studied at Cambridge and became a country rector not far from Newton's birthplace. She married again, her new husband also a rector but a wealthy one. She had three more children but no use for her firstborn, who was left in the care of his grandmother. Her second husband then died, and she and her new children returned to Woolsthorpe—but now Isaac, at the age of ten, was sent off to a school several miles away. After he was abandoned by his mother again, solitariness became his lifelong habit, whether because it suited him or because he could conceive of nothing else we shall never know.

In 1661, he enrolled at Cambridge, a poor student left to make his own way despite his mother's now considerable wealth. Aristotelianism was still the backbone of a Cambridge education, but new scholars—Copernicus, Kepler, Galileo—were making themselves known. Newton found their books at the university library, devoured them, and wanted more. He found the old works of Euclid and discovered the new works of René Descartes, who provided an important innovation: Descartes imagined the path of a moving object as a line drawn between two perpendicular axes, a graph of distance versus time. Cartesian coordinates, they are still called. But Descartes was a philosopher more than a mathematician, and he could not take the analysis any further. He also shared the traditional philosopher's dislike of algebra.

After three years Newton took his examinations and became a scholar at the university. He was helped by Isaac Barrow, Cambridge's first Lucasian professor of mathematics, who saw something out of the ordinary in his unworldly student. But then the plague struck

England, as it did periodically, and reached Cambridge. Scholars and students dispersed. The cause of plague was unknown, but everyone knew that once it hit a town of any size and density, it ran through like fire. Newton retreated with his books to Woolsthorpe and there, at the age of twenty-two, he invented what became the essential tool of mathematical science.

Puzzling over the nature of continuous but variable motion, Newton could see, as Galileo had sensed, that it all came down to the mathematically vexing issue of infinitesimally tiny distances traversed in infinitesimally tiny periods of time. Newton first attacked the problem of calculating the total distance a body travels in a finite amount of time, when you know its velocity at each moment. Draw a graph, à la Descartes, showing velocity on the vertical scale and time horizontally. Let it wiggle up and down as it may, the only limitation being that it must be a continuous curve (velocity cannot jump abruptly from one value to another but must pass through all the values in between).

At each point on the line, Newton said, the body travels an infinitesimal distance equal to the velocity multiplied by an infinitesimal amount of time. Graphically, this distance is the area contained by a

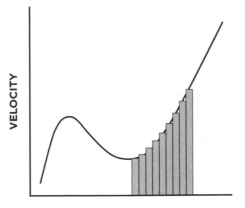

VELOCITY

TIME

rectangular sliver with the height of the line. The total distance traveled is then the sum of all these little areas. To get an ever more accurate evaluation of this distance, let the rectangular slivers become thinner and thinner, and so equate a finite amount of time to a larger and larger number of such slivers.

This trick is closely related to what ancient Greek mathematicians called the *method of exhaustion*. Archimedes had used it to calculate the area of a circle: take a circle and draw a square within it; the area of the square is clearly smaller than the area of the circle. Now replace the square with a pentagon. Its area is smaller than the circle's but larger than the square's. Replace the pentagon with a hexagon, the hexagon with a heptagon, and thus proceed ad infinitum. Each polygon's area is larger than its predecessor's but always smaller than that of the circle itself. In short, as you go to polygons with ever increasing numbers of edges, you get closer and closer to the area of the circle.

Newton was doing the same thing to find the area under a section of the velocity graph by filling it with an increasing number of narrow rectangles. But there was a huge complication. A circle is a well-defined shape, so it was straightforward (if tedious) for Archimedes to figure out the size and area of a polygon that just fit within it. Newton, by contrast, wanted to allow the shape of his graph to be anything he wanted, as smooth or wiggly as he cared to make it. To jump to modern language, we say that velocity is a function of time: for each value of the time, there is a corresponding value for the velocity.

Newton showed how to let the number of infinitesimally thin rectangles under the velocity curve grow without limit, toward infinity, provided that they get thinner as the number of them grows in such a way that the total time elapsed remains fixed. And he showed how to add up the areas of this almost infinite number of almost infinitesimally thin rectangles to obtain a finite and consistent answer. He showed how to calculate the area under the curve to get the distance traveled. In modern language, he had shown how to calculate the *integral* of the velocity function.

Newton then addressed the inverse problem. If instead of a graph of velocity we have a graph of distance traveled as a function of time, then at any point on that curve, the velocity is the ratio of the infinitesimal distance the object moves to the infinitesimal time the object takes to move it. It is the slope of the curve, the tangent line to it. Newton showed how to calculate its value by imagining that the infinitesimal distances and times become ever smaller and smaller, while their ratio converges on a well-defined value. This value, the object's velocity at any moment, is the *differential,* as we now say, of the function of distance versus time.

In short, Newton invented differential and integral calculus. I learned the beginnings of calculus, as best I can recall, when I was about fifteen or sixteen years old, and even now I can remember the ingenuity, the cleverness, the sheer elegance of its procedures. The heart of calculus is its ability to make rigorous and consistent sense of infinities and infinitesimals. It tells why Zeno's paradox is mistaken. The frog may take an infinite number of jumps, each one tending toward an infinitesimal length, but the sum of an infinite number of infinitesimals can be finite.

If Newton had invented calculus and done nothing else, he would still be a remarkable figure in the history of mathematics and science. But he did much more. He formulated the essential laws of motion, putting into exact terms ideas that Galileo had intuited but could not phrase precisely. First, an object at rest or moving at constant speed continues in that state unless disturbed by a force—this defines the property we call inertia, the tendency of a body to keep doing whatever it's doing unless disturbed. The second law says that what causes a change in the body's state of motion—that is, an acceleration—is a force; specifically, the acceleration is the magnitude of the force divided by the object's mass (the heavier the object, the harder you have to shove it; and if you push back on a body moving toward you, you reduce its velocity, thus causing a negative acceleration, or deceleration). The second law defines quantitatively what we mean by a force, a word that before then had had only a vague connotation. Third, each force generates an equal and opposite reaction—so that when you push on a shopping cart, you feel the cart pushing back on you, creating a force that you feel in your hands and arms.

That was still not all: Newton devised the law of gravity and proved that the orbits of the planets around the sun must be ellipses (the circle being the special case of an ellipse with zero ellipticity). By means of his gravitational law, he explained the tides (a problem Galileo had wrestled with throughout his life, failing to find a convincing solution). He showed that light comes in the colors of the rainbow, and that white light is all those colors equally mixed. He studied reflection and refraction, and invented (and built, with his own hands) a reflecting telescope free of the chromatic distortions that lenses always have, because they refract different colors of light by slightly different amounts.

The story of how Newton's inventions became known to the world is as strange as the man himself. In his early years, his mentor Isaac Barrow told the scholars who created the Royal Society in London that he knew of a young man who was making astonishing discoveries. Newton was cajoled into publishing some of his ideas in teasingly limited ways, always hinting that there were other great things he could do and had done, but saying no more about them. These publications led to arguments, in the nascent scientific press, about whether he was right, whether he could prove what he said, whether he had "borrowed" crucial ideas from others. Newton's response was to write viciously scathing rebuttals, calling his critics fools and nitwits, then retreat into magisterial silence. And after his early inventions in science, little of which he revealed, he proceeded to spend years and years attempting to turn base metals into gold, the old alchemical dream, and poring minutely over the Bible to establish a thorough chronology of the ancient kingdoms of the Middle East.

Newton's secretiveness caused him great woe. The German philosopher Gottfried Leibniz independently worked out the mathematical principles of the calculus after Newton, but Leibniz published his work promptly. When Newton's achievements eventually came to light, Leibniz's followers accused him of plagiarism, which led Newton to release some early notes he had sent to Leibniz and suggest that, au contraire, Leibniz had got the idea from him. Same thing with the inverse square law of gravity: the idea had been floating around for a while, and Newton was one of several thinkers to seize on it. But only Newton was able to figure out what it implied, mathematically, for the orbits of the planets. And again, when Newton's work became known, there were accusations of intellectual theft and plagiarism.

That Newton's work emerged at all was almost happenstance.[2] In 1684, the astronomer Edmond Halley had been discussing with others whether an inverse square law of gravity meant that planetary orbits must be ellipses. They suspected it was true but could not prove it. Some time later Halley was visiting Newton and put the question to him. Yes, said Newton, without hesitation, an inverse square law required orbits to be elliptical. Halley asked how he could say so. Because, said Newton, he had worked it out a long time ago. Halley implored Newton to write up these proofs and settle the debate— and, by the way, establish his primacy before other pretenders came along. A few months later Newton sent Halley a small pamphlet with the details of his proof, which went both ways: he showed that elliptical orbits imply an inverse square law of gravity, and also that an inverse square law implies that orbits must be either elliptical, circular (a special case of ellipticity), or hyberbolic, the last applying to a body such as a comet that speeds in from outside the solar system with too much energy to be captured but arcs around the sun before speeding out again.

Halley wanted to publish this work, but Newton wanted to wait— not, this time, because of his pathological secrecy, but because Halley's visit had set him off on a quest to go back through all his old work on motion and mechanics, revise and expand it, work out better proofs of old theorems, establish new theorems, and generally flesh out his collection of notes into a complete work of natural philosophy unlike anything that had been seen before. The task took two years of concentrated effort, during which Newton was occasionally seen wandering around Cambridge, consumed in thought, but was mostly absent from public view, holed up in his rooms. The outcome, in 1686 and 1687, was the publication of the three momentous volumes of the *Philosophiae Naturalis Principia Mathematica,* the *Mathematical Principles of Natural Philosophy.* These books are the founding documents of modern physics. Some considerable time later, in 1704, Newton was persuaded to publish (in English rather than Latin) his *Opticks,* a book more experimental than theoretical in nature, containing his ideas on the behavior and properties of light, plus sundry observations on the natural world and a list of questions demanding further investigation.

In the *Principia Mathematica* and *Opticks,* Newton "made knowledge a thing of substance: quantitative and exact. He established principles, and they are called his laws," as James Gleick nicely puts it.[3] Whole libraries have been written on the origin of these works and their standing as the entrance gates to the modern era of science. I have emphasized in this chapter his invention of calculus because it is this mathematical tool that finally cleared away the philosophical fog obscuring the concept of motion. It is what enabled Newton to make mathematically precise statements about objects moving with variable speed and variable acceleration and to formulate laws that applied to any form of motion, not just the uniform speeds and constant accelerations that had been the limits of Galileo's insight.

Ultimately, Newton's work displaced geometry from its reigning position as the perfected form of mathematics and enshrined algebra as the practical mathematics for science and scientific laws. But here again Newton caused difficulties that slowed full appreciation of his work. When I was about sixteen, and had learned the rudiments of calculus and Newton's laws of motion, I thought it would be interesting to go back to the source. My school library had a copy of the *Principia Mathematica,* translated from Newton's Latin into English, and I opened it, expecting to see the principles and equations I had recently learned. But the thing was incomprehensible! Instead of equations, the pages were adorned with elaborate but inscrutable geometrical diagrams accompanied by proofs that were likewise expressed in geometrical form, as if Newton had gone out of his way to transform his invention of calculus into something that Archimedes and Pythagoras might have recognized. Which was the case, more or less. Despite Newton's great innovations, the old language of geometry was still the time-honored and philosophically respectable way to present mathematical truth, and the inventor of calculus himself was unable to unburden himself immediately of the old tradition.

Leibniz, in his presentation of calculus, used algebra without shame, and his method and notations became standard. Newton's insistence on writing in the combined languages of Latin and geometry only added to the impenetrability of the *Principia*. Later, in *Opticks,* Newton turned to English and set out theorems in a recognizably modern form, with proofs conveyed by calculus in the form of equations. Even then, Newton stuck to his own notation of calculus, which was cumbersome compared to the system Leibniz had by that time made public. At Cambridge University in the 1970s there were many lecturers who continued to use Newton's notation, in deference to the great man but not to the advantage of undergraduates. In fact, an excessive devotion to Newton inhibited the further development of calculus in England, and it was Continental mathematicians and physicists who used it to analyze and solve ever more complex cases of motion.

But that's by the way. The combination of calculus and clearly enunciated laws of mechanics and gravity created what we can fairly call the Newtonian view of the natural world, governed by measurable and observable effects following from quantitative causes. And calculus wasn't just applicable to motion. Any physical phenomenon that changes with time—a rising or falling temperature, an increasing or decreasing pressure—could be depicted with this new mathematics. The vexatious issues of motion and change that had defeated the ancient Greek philosophers were now brought into the fold of mathematically manageable problems. Calculus became the essential tool of modern science. It made motion and change and flow the basic elements of theorizing, an enormous contrast to the older philosophical emphasis on constancy. The ancients accepted planets moving in circles, of course, but only because the circle was seen as the most constant and therefore most perfect form of motion. Newton made any form of motion or change fair game for the mathematical thinker. If Galileo created the idea of science in its modern form, it was Newton who made it powerful, analytical, and universal.

7

The Language of Mathematics

With the invention of calculus, modern science was off to the races. Gone at last was the fixation on geometry as the purest, most ideal form of mathematics, and gone, too, was the principle that geometrical forms were the building blocks of nature. Instead, mechanics became the exemplary science. Calculus made it possible to analyze motion and force in completely general terms. An object could move with ever-changing speed and direction, subject to multiple and variable forces, but Newton's laws of motion, coupled with calculus, allowed an eager scientist to capture all the relevant causes and effects in precise mathematics.

The study of motion began with the paths of planets and cannonballs, but the versatility of Newtonian mechanics soon widened the compass of what scientists could do. I am engaging in a little ahistoricity by calling researchers in the time of Newton "scientists"—that word wasn't invented until the nineteenth century. Investigators of nature at the time regarded themselves as philosophers, more specifically natural philosophers. But in deed and thought they were scientists as we now recognize that species, observing, measuring, theorizing, and calculating in a recognizably modern way.

Galileo was aware that the path of a falling stone or a zooming cannonball would not, in the real world, be precisely what his geometrical methods said it would be, because the air through which a stone or cannonball passes exerts a little resistance on moving bodies. But he had no way to incorporate the effect of air resistance into his thinking. Intuitively he would certainly have been able to see that the cannonball would fall short of its predicted target, but by how much he could not say. In Newtonian terms, however, friction from air is another force that can be added to the equations of motion. Early scientists would not have had much of a handle on how big that force would be, but they could at least make a start on estimating its significance.

Similarly, Newtonian theory made it obvious that the path of a planet orbiting the sun is influenced to a tiny degree by the gravity of the other planets. Calculating the orbit of one body orbiting another is straightforward, but adding a third creates the notoriously difficult three-body problem, a mathematical challenge that Newton understood to be far beyond his abilities to resolve. Still, when you have one very big object, the sun, and two or more much smaller ones, the planets, it becomes possible to estimate approximately one planet's gravitational influence on another's motion. In the early eighteenth century, a number of astronomers studied the orbit of Uranus and deduced that its orbit was being slightly disturbed by another planet, farther out. Neptune was duly discovered in 1846, close to its predicted position.

Calculus and mechanics clarified many other archetypal problems. Galileo had thought about a wooden beam, embedded firmly at one end in a wall and loaded with a weight at the other. How much would it bend and what shape would it take up? He made remarkable progress with his geometrical analysis, but a complete picture of the physics was beyond him. Calculus helped. When such a beam bends, its departure from horizontal grows incrementally with distance from the fixed end. The shape of the bent beam can be drawn on a graph, Cartesian-style. The slope of that curve—the differential of its displacement, in the language of calculus—tells you about the stress in the beam at that point. The total displacement of the beam at its end is the sum of all the infinitesimal displacements along its length—a problem in integral calculus.

It's important to see how empirical knowledge plays into this kind of physics. Beams of different material will bend by different amounts. Mathematical analysis shows that this difference can be related to a certain strength modulus, a number that puts in quantitative terms the stiffness or flexibility of a given material. Doing the experiment and making the measurements feeds into the mathematical framework to determine a basic property of the material. The novel point was that estimating the strength modulus of some material relied not at all on knowing why one material was stiff and another floppy. What difference in their internal composition made them so? It didn't matter. The number measured something about the material, regardless of what the material actually consisted of. On that score, nobody in the early days of modern science had much of an idea. But to depict mathematically how real materials bend, they didn't need to.

This was a huge contrast with prior thinking. In old style philosophy, what counted was knowing the innermost structure of a material—brick or brass or oak. Only then could you hope to understand how it behaved. Science turned that kind of thinking around. The susceptibility of a given material to bending could be characterized by a number that described the *empirical* nature of that material. Why brass bent more easily than stiff oak was a problem for another day. What mattered was that an engineer or architect could select brass or wood or some other material on account of its known properties, and could reliably calculate how it would perform in one kind of construction or another.

The old philosophers would have been appalled. Where were the grand ideas about the constitution of matter? Where were the all-encompassing hypotheses of the nature of physical reality? Working out how a beam would bend was the philosophical equivalent of manual labor, mere craftsmanship instead of profound cogitation.

But that was exactly why modern science worked. It focused on the practical and doable tasks at hand. Pragmatism was its guiding principle.

The point of pragmatism is to get useful answers. Putting Newtonian science into practice meant writing down equations to capture the physics of some given situation and then solving those equations. It's worth pausing for a moment to delve into what's entailed here. How does a scientist pose a physics problem in mathematical terms and then find—or not—a solution?

Isaac Newton told a surprised Edmond Halley that he had long ago proved that planets orbiting the sun must follow elliptical paths. To see how Newton did this, think of a planet at a specific point in space with some specific velocity. The sun's gravitational force on the planet depends only on the distance between the two bodies. That force exerts an acceleration that changes the planet's velocity, so that the planet's position and velocity will be infinitesimally different an infinitesimal time in the future. What we've constructed is that workhorse of Newtonian science, a differential equation: for a given set of conditions, it tells you what happens a moment later; and when that moment has elapsed, the same differential equation tells you what happens after the next moment has elapsed.

The solution to this differential equation is the planet's trajectory. From an arbitrary starting condition, the planet might do any number of things. It might fly off into distant space. It might spiral down into the sun. It might loop around the sun a few times on a complicated path before flying off or spiraling down. What you are looking for, though, is a special solution of the governing equation corresponding to a path that loops back on itself and returns to its starting point. It will then follow the same path and come back again, ad infinitum. This is a stable, everlasting orbit, the desired mathematical solution to the problem Halley posed to Newton. The only paths that work are ellipses, which give you the planet's position and velocity at any point in its orbit. It goes slower when the planet is farther from the sun, faster when nearer, just as Kepler found in this second law of (observed) planetary motion.

I apologize to readers who are familiar with this sort of procedure. But it has been my experience that phrases like *finding an equation* or *solving an equation* are highly mystifying to those who are not in the know. I don't mean to teach such readers how to construct and solve equations, but I want to make clear what the process is about. I want to make clear how the mathematics of the equation embodies what we know about the physics, and how the solution accordingly represents motion that is consistent with the constraints set by the laws of physics. This is how math, in the age of Newton, serves science: it captures natural laws and tells us what kind of behavior and phenomena are in accord with those laws and which are not.

In the case of planetary orbits, the equations and the solutions are not too challenging. But that's by no means always the case. For other physical phenomena, you may be able to write down equations that precisely describe what is going on, but you may not easily be able to solve those equations. They are too difficult.

Okay, but what does *that* mean? It's a subtle matter. In the case of orbits around the sun, it's an easy exercise to show that when a planet moves on an elliptical path, the differential equation for its motion is satisfied. What's more, an ellipse is defined with simple algebra, so that even in the seventeenth century it was no big deal for an astronomer to work out planetary positions with pen and paper. The task was tedious, to be sure, but not intellectually demanding; it was merely a matter of following instructions, as with a recipe.

Another case of simple motion that turns up frequently is the sine curve, characteristic of waves and oscillations. A weight suspended on a spring and bobbing up and down; a taut violin string plucked to set it vibrating. Sine waves are ubiquitous because they are the solution to an elementary differential equation, one that applies whenever the force returning an object to a neutral position is proportional to how far the object has moved. A weight will sit motionless on a spring at its balanced position, but tug it down a little and the spring pulls back. Similarly, a curved violin string wants to go back to being a straight line. In both cases, the force returning the weight or the string to its neutral position is proportional to how far it has strayed. A sine wave oscillation results.

The sine function originated in trigonometry as one of the basic properties of right-angled triangles, and the wave it generates can be seen as the variation in height of a triangle that rotates at constant speed within a circle, with one vertex at the center and another moving around the circumference. Hence the deep connection between sine waves and circular motion, which is why they turn up so often in physics.

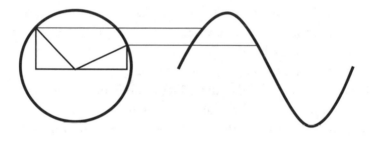

Unlike an ellipse, a sine curve is not easy to calculate with pen and paper. These days, of course, you can just touch an icon on the calculator app on your phone, or if you are programming a computer you put $y = \sin(x)$ or something similar in the code. When I was a teenager, just before the appearance of electronic calculators that could fit in your pocket, we had slim booklets that tabulated the value of the sine function to four or six or more decimal places. But where did those numbers come from?

You could try to tabulate the values of the sine function by drawing triangles and measuring the appropriate ratio, but even the most accurate draftsman will at best get a couple of reliable decimal places from such a procedure. However, there's an algebraic way of working out the sine function, in the form of an infinite series. Here's the equation, a very straightforward one:

$$\sin(x) = x - \frac{x^3}{3!} + \frac{x^5}{5!} - \frac{x^7}{7!} + \cdots$$

You know that x^3 is simply x to the third power, meaning x times x times x. You may not know that 3! is three factorial, meaning 3 times 2 times 1, equaling 6. Similarly with the other terms in the series, which goes on forever. The terms get steadily smaller, however, so that if your task is to calculate the value of the sine function to a certain number of decimal places, you can put down your pencil once the value of the next term is small enough that it makes no difference to your answer.

In the days before electronic calculators, human calculators would devote many days and lamplit nights to working out values for the sine function and publishing the results in books containing lengthy tables of numbers—rows and columns of numbers, one after the other, not unlike the tax tables you can find in the instructions for IRS Form 1040. The reason they did this is because of the ubiquity of the sine function, which turns up again and again in the solution of differential equations for motion.

This is why a mathematical physicist who finds that sines (perhaps in combination with other trigonometric functions, which have broadly similar properties) pop out of a differential equation will say they have "solved" the equations. An answer has been found in familiar and standard terms that you can look up in a book. Every scientist knows what a sine wave looks like, that graceful, endless up and down. It is intuitive because it is so familiar. It is a known quantity, a valued friend.

Still, for most physics problems, sine waves don't show up. When I was an undergraduate in the 1970s, I took a course with the title, if I remember correctly, Mathematical Methods for Theoretical Physics. The bulk of the course consisted of lessons in how to solve differential equations, drawing on the achievements of applied mathematicians of the eighteenth and nineteenth centuries. We learned all about the common trigonometric functions and their properties, of course, but that was kids' stuff. All kinds of other functions, adorned with the names of French and German and British and Russian mathematicians, were put before us.

Here's an example. A plucked violin string oscillates, as I said, in the form of a sine wave. But what about a circular drumskin, made of some taut material and fixed firmly around its circumference? When you hit it with a stick, what kind of oscillations do you set off? It's fairly straightforward to write the differential equation that tells you how strongly any point on the drumskin wants to return to its neutral position. The problem, as always, is getting a solution. Daniel Bernoulli, perhaps the greatest of the extended Swiss clan of Bernoullis, first studied this equation in the eighteenth century, but it was the German mathematician Friedrich Wilhelm Bessel who, a little later, organized and codified its solutions, which are now known as Bessel functions. Like sine waves, they can be expressed as infinite sums of powers of x, so that a hardworking calculator (a human one) can work out their values and tabulate the results in fat handbooks. Bessel functions are not as commonplace as the trigonometric functions, but they crop up often enough that it was worth being able to go to your library (in the precomputer age, I mean) and find a table of the Bessel functions rather than having to do all the arithmetical drudgery yourself.

Bessel functions describe all possible vibrations of a drumskin. The simplest is when the whole drumskin goes from convex to concave and back again, the center moving up and down rhythmically. There's also a Bessel function corresponding to when one side of the skin is going up while the other is going down, with a diametrical line separating the two halves staying still. Then you have a solution when the drumskin vibrates in quadrants, with adjacent areas moving in opposite directions. Bessel functions come in an infinite number of varieties, and any oscillation of the drumskin can be written mathematically as a combination of Bessel functions. This is exactly comparable to the way that any vibration of a violin string is a combination of its fundamental modes, which are all sine waves of different frequencies.

If you were in a line of work where Bessel functions cropped up regularly, they would become your friends, as familiar to you as sine waves. If you solved an equation and found that your answer could be written in terms of Bessel functions, you would think, aha, I really have *solved* this equation. I've nailed this thing! I know what it does and how it behaves. I have a mental picture that lets me see the physics.

The toolbox of nineteenth-century mathematical physics, like the workshop of a master watchmaker, was stuffed with ingenious gadgets. Legendre polynomials, Chebyshev polynomials, Hermite polynomials; Fourier and Laplace transforms; hyperbolic trig functions and elliptical integrals; and, most splendid of all in my recollection, the method of steepest descent. It doesn't matter what all these things are, except that they were invented by mathematicians or mathematically inclined physicists for the purpose of solving equations. There is no single universal method for solving differential equations. With experience, you learn tricks and techniques. You find ways to transform a gnarly looking equation into something a little more familiar, a little closer to your personal encyclopedia of known forms.

Through familiarity with the methods of solving equations the physicists of the classical era gained familiarity with the mechanisms of physics itself. Although many of these tools had properties of great interest to pure mathematicians, it was their application that entranced the working physicist. Understanding the mathematical properties of equations and their solutions allowed physicists to think that they understood how nature itself behaved.

As it happened, I learned these old mathematical methods just before they passed into antiquity. As undergraduates, we had access to the university's mainframe computer, but only for special occasions. I remember one lecturer writing a complicated differential equation on the blackboard and asking for ideas on how to solve it. Various brave souls raised a hand and proposed one thing or another, and the lecturer wrote down a numbered list of our suggestions. Then someone had the temerity to suggest using a computer to solve the equation. The lecturer, an elderly gent, scowled and marched to the other end of the blackboard, where he wrote "99: use the computer."

There was a real feeling, I remember, that solving an equation with a computer wasn't really solving it. Sure, the computer could churn through the numbers and spit out answers, but how would you know what those answers really meant? If your answer, as in the old days, was a collection of sine waves or Bessel functions or the like, you could claim to have a mental picture of the solution, because you had some familiarity with how sines and Bessel functions behaved. But a long list of numbers zapped out by a computer? For a time, no one quite knew what to make of such "solutions."

That's ancient history, of course. Nowadays we have computers on our desktops that are more powerful than the mainframes of my undergraduate days, and they are often the first recourse in solving an equation, not the ninety-ninth. Undergraduates today take classes in how to perform computer calculations and simulations and, alas, know little of the wonders of elliptical functions and Chebyshev polynomials. One of the keys to gaining physical insight from computer-generated solutions came from the ability to display such a solution as a graph on a screen rather than a printed list of numbers. In the early days you might need hours to generate a stack of printed sheets listing your solution in rows and columns of numbers. It's hard to look at that kind of output and gain any intuitive sense of what the answer means. Nowadays, though, you can often generate a solution more or less instantly, display it as a graph on a screen, tweak the parameters a little, and display the modified solution side by side with the earlier one. You can see in real time how fiddling with the physical characteristics of a phenomenon alters its behavior. Even better, computers can generate three-dimensional maps in a multitude of colors, the kind of thing you see every day on the weather forecast to illustrate how an impending storm is moving into your area and where it is likely to be most severe. For modern scientists, familiarity with computer-generated solutions has largely supplanted the familiarity that scientists in the past gained from their comfort with Bessel functions and their numerous cousins.

My purpose in telling you all this is to bring home the idea that "solving an equation" is a somewhat elusive notion. It depends on the tools that a mathematical scientist has to hand, and those tools change from one era to the next. To a nineteenth-century scientist, a solution was a collection of known mathematical functions written on a piece of paper—a recipe for finding out what was going on in whatever system you were studying. Today, a solution is a complex computer-generated image. What hasn't changed, though, is Galileo's dictum that mathematics is the language of the universe. Fluency in that language is what allows insight into natural behavior.

But—and I want to stress the point again—it is indeed a language, a means of translating the workings of nature into symbols or numbers or graphs or computer-made movies. These mathematical tools portray the universe; they don't create it. It's the underlying laws of nature that do that.

8

The Limits of Pragmatism

Newtonian science began with mechanics but it didn't stop there. A thorough understanding of the wave equation (the one that generates sine curves in its solutions) enabled physicists to make some progress on the nature of light. Newton believed light was a stream of particles, but as time passed that became a minority opinion. Mathematical investigation of the properties of waves suggested that they were a better fit for the observed characteristics of light. For example, the famous Swiss mathematician Leonhard Euler proved in 1746 that if the different colors of light corresponded to different wavelengths, they would refract differently, explaining why telescopes produced colored fringes at the edges of planet images and the like.

But if light was a wave, a big question arose: waves in what or of what? Waves on the sea are easy to grasp: the surface of the water moves visibly up and down. Waves in the air, constituting sound, are a little more abstruse: we can't see them, but we can hear them, and the behavior of sound can be captured in the mathematics of waves of pressure, alternately higher and lower, traveling through the air. But for all that light was itself highly evident—it's what our eyes detect, after all—its nature remained inscrutable. If light was a wave, what was doing the oscillating? What medium were the waves traveling through?

It was a sign of the power of mathematical physics that theorists could construct arguments about the behavior of light and design experiments to test them without having the slightest idea, in fundamental terms, of what they were experimenting on. The pragmatic attitude prevailed. Worry about what a thing *does,* not what it *is.* The scientist of the Newtonian era could forge a new understanding of the phenomena of light, backed up by theory and experiment, without needing to think too much about what light actually was.

In this way, mathematical physics became an exploratory tool of unprecedented power. But that power had its limits. The centuries after Newton saw enormous progress on many fronts. It was the period in which much of the basic science of the world around us was built. I want to focus, though, on two illustrative examples, heat and electromagnetism, that depict the shortcomings as well as the power of classical mathematical physics.

Heat is both obvious and mysterious. We sense the difference between hot and cold as easily as we tell light from dark. Simple thermometers, based on the observation that gases expand when heated, go back a long way. Galileo made an air thermometer, but it didn't work very well because it was also affected by atmospheric pressure, which he didn't know about. The first accurate modern thermometer was invented in 1714, by Daniel Fahrenheit. It used a column of mercury and had a stability and reliability far superior to any previous thermometer. Fahrenheit defined the zero of his scale as the lowest temperature he could reach by mixing salt and ice and then, for enigmatic reasons, defined 32° F as the temperature of pure water at its freezing point and 96° F as the normal temperature of the human body (his own, one presumes). Later, the boiling point of water was set at 212° F, 180 degrees from the freezing point. The specific numbers are not important. Fahrenheit's thermometer made it possible to reliably associate numbers with the previously qualitative concept of hotness and coldness. It made temperature measurable and quantitative, even though it provided no enlightenment as to the physical nature of the thing measured.

Another aspect of heat proved more puzzling still. That it takes longer to boil a big pot of water than a small one must have been known since ancient times. The temperature of the water at the end is the same, but it seems that the larger amount contains a greater quantity of *something*. This *something*, moreover, evidently had the ability to flow from one place to another. Place a hot piece of iron next to a cold one, and we all know what will happen: the hot one cools down and the cool one heats up. Put a small gap between the two chunks of metal, and the same thing will happen, but much more slowly. A lump of hot iron placed in the open air will likewise cool down, its *something* evidently dispersing into the atmosphere.

Along with degree of heat, which we call temperature, there was also a quantity, an amount of something, that was simply called heat. The idea of heat as some sort of intangible fluid, flowing through the crevices of different materials at different rates, gained traction. In 1822, Jean-Baptiste-Joseph Fourier published one of the great works of the Newtonian era, his *Théorie analytique de la chaleur* (*The Analytical Theory of Heat*). In it, he brought to bear a variety of novel mathematical techniques to portray how heat flowed through materials; notable among these tools was the Fourier series, a way of expressing complex mathematical functions as the sums of sine waves of different frequencies. Fourier's analysis was an exquisite exercise in the depiction of a physical phenomenon despite the fact that it neither required nor yielded any insight into the fundamental nature of that phenomenon. Pragmatism again: to analyze how heat moves, you don't need to worry about what it is.

The scientific investigation of heat became more urgent with the emergence, late in the seventeenth century, of the steam engine. The inventors of the steam engine were ingenious men with little or no interest in the science of heat, but they knew that heat could transform a small volume of water into a large volume of steam, and they transformed that expansion into mechanical motion by making the steam push on a piston. The first use of steam engines was to power pumps that raised water out of mine shafts. Soon afterward, engines were made to haul heavy loads along tracks. In time, the mechanical engine put horses and donkeys out of business, which in the long run was greatly to the benefit of those ill-used animals.

The activities of the inventors quickly caught the attention of more scientifically minded thinkers, who began to wonder about this curious ability of the intangible fluid called heat to drive very tangible machinery. The science that thus emerged was thermodynamics—the mechanics and motion, Newtonian-style, of heat. The tale is exceedingly complicated. The big sticking point in the early days was the belief that heat was a substance in its own right, and that although it could flow and change temperature, it couldn't change into anything else. A great early achievement was a remarkable 1824 booklet by Sadi Carnot with the title *Réflexions sur la puissance motrice du feu et sur les machines propres à développer cette puissance* (*Reflections on the Motive Power of Fire and on Machines Able to Develop That Power*). Carnot imagined heat flowing through a steam engine in somewhat the same way that water flows through a water mill. Water, flowing downhill, transfers some of its energy to the mill wheel. Similarly, heat, flowing from high temperature to low, causes a piston to move. Although this analogy was wrong, Carnot came to some correct conclusions, most notably that it was impossible to turn all the heat into mechanical motion. There would always be some inefficiency, some heat left over. This observation is the root of the second law of thermodynamics, the one about entropy always increasing.

Not only was Carnot mistaken about the nature of heat, his analysis rested on some slightly dodgy mathematical sleight of hand (in essence, he made assumptions in some parts of the argument that he forgot about or ignored in other places).[1] The fact that he was able to get to the right answer with both incorrect physics and flawed mathematics speaks to his intuitive sense about how physics must work. To be a scientist is not to be a mindless drudge plodding through strictly logical propositions and conclusions. A leap of the imagination is always present in the greatest of achievements. Often, it is only later that logical rigor and correct mathematics catch up. That was certainly the case with thermodynamics. It was not until the middle of the nineteenth century that scientists realized that energy comes in many forms but can be neither created nor destroyed. This is the principle of the conservation of energy, also known as the first law of thermodynamics. Only then was heat itself recognized as one of the forms that energy could take, and that in a steam engine (contrary to what Carnot supposed) some part of the heat energy is actually converted into energy of motion.

What became known as the classical laws of thermodynamics finally became clear around the mid-1860s, when the second law was presented in a strictly defined way. Thermodynamics, it was then apparent, was much more than the science of heat and engines. It was the science of energy itself, and its laws applied to any physical or chemical or even biological transformation. Animals gain energy from the food they eat; they expend it running their metabolisms and running after prey or from predators.

At this point, thermodynamics was a full and precise science, captured in rigorous mathematics, but it still hadn't answered one question: What, actually, is heat? Yes, it's a form of energy, and that explains its behavior, but it doesn't explain its nature.

The answer was not long in coming. A few scientists, notably Ludwig Boltzmann in Austria and James Clerk Maxwell in Great Britain, had begun to take seriously the ancient notion that all matter was made of atoms—minute objects that came, somehow, in different forms or combined in different ways to create the enormous variety of stuff in the world around us. Specifically, Boltzmann and Maxwell pictured a volume of gas as an enormous number of tiny, hard masses speeding about and colliding with each other in mostly empty space.

These atoms pleaded allegiance to the Newtonian laws of mechanics: they had velocities and directions, carried momentum and kinetic energy, bounced off one another in predictable ways, and so on. The kinetic theory of heat yielded some highly satisfactory conclusions. For one, the pressure exerted by a gas on the walls of its container was the macroscopic manifestation of numerous microscopic collisions of atoms with the wall. The greater the average energy of the atoms, the greater the pressure.

That insight was intimately tied to a far greater one. If faster atoms mean higher pressure, and higher pressure comes from hotter gases, then it was not so hard to prove that the kind of energy we call the heat of a gas is nothing more mysterious than the total kinetic energy of all its atoms. Heat, that long mysterious substance, that enigmatic quasi fluid, was revealed not merely as a form of energy but as one of its most elementary forms, the energy of motion. This was a huge triumph for Newtonian thinking.

There was, at the time, a small element of dissent from this triumphal note. The Austrian physicist Ernst Mach (of Mach number fame) vehemently opposed the kinetic explanation of heat on the grounds that atoms were entirely hypothetical. He had a point: there was no direct evidence for the proposition that all matter is made of tiny particles of unknown composition, so wasn't it irresponsible to propound far-reaching theories on unproven assumptions? Mach's attitude was that physicists should stick to what they know and what they can measure or demonstrate directly. He liked the old version of thermodynamics, in which heat obeyed strict mathematical rules but in which the nature of heat was left undetermined. Physicists, said Mach, should not pretend to know about things they can't directly test.

On the other hand, science makes progress by coming up with hypotheses and seeing how they work out. Are they effective? Do they provide satisfying explanations? Do they make sense (I mean, do they make sense in an intuitive rather than strictly logical way;

science is never a matter of strict logic and deduction)? The atomic hypothesis passed those tests. It proved valuable in other areas of physics and, of course, in chemistry. It was some time before the existence of atoms was directly demonstrated, and only with the advent of nuclear physics and quantum theory did it emerge that atoms themselves are complicated structures.

Mach had his adherents, but his extreme view amounted to an attempt to bottle up natural curiosity by saying that scientists shouldn't dig beneath the surface. An accurate, if superficial, mathematical description should suffice, he claimed. But this was far too limiting a philosophy and his views, never widely popular, quickly faded.[2]

Still, Mach touched on an important issue. By the late nineteenth century, scientists were comfortably accustomed to depicting nature in mathematical terms, but how far can mathematics alone represent what we might call reality? The atomic explanation of heat proved appealing precisely because it gave the mathematical laws of thermodynamics an appealing physical foundation. Tiny atoms zooming about and crashing into one another: it all made sense. And it made sense especially because it pictured the microscopic world in the same terms as the world at large. It made Newtonian mechanics the ruling theory of what we can't see as well as of what we can.

It was around this time, in the latter half of the nineteenth century, that the trope of the natural world as a machine gained popularity. Factories were commonplace, the engine of the new industrial economy. Trains crisscrossed the land. Wealthy people could afford intricately designed watches and other mechanical knickknacks. Machinery was everywhere, and if Newtonian physics was the master theory, must not the universe as a whole itself be a machine, a system of moving parts designed with exacting precision by the Creator?

This was not an entirely new idea. In a much quoted observation, the marquis de Laplace, a noted French mathematician, wrote that:

An intellect which at a certain moment would know all forces that set nature in motion, and all positions of all items of which nature is composed, if this intellect were also vast enough to submit these data to analysis, it would embrace in a single formula the movements of the greatest bodies of the universe and those of the tiniest atom; for such an intellect nothing would be uncertain and the future just like the past would be present before its eyes.[3]

This was written in 1814, in Laplace's *Essai philosophique sur les probabilités* (*Philosophical Essay on Probabilities*), when Newtonian science was in the ascendant but far from its peak. Laplace was making a leap of faith in asserting that all nature would eventually come under the rule of Newtonian-style laws, mathematical and inerrant; he was expressing confidence in a system yet to come. But his remark encapsulated what would become the credo of the mechanical view of the natural world: a system of forces acting on "all items of which nature is composed." It was a considerable supposition, in 1814, that nature would indeed turn out to be built that way.

The trope of nature as machine, of the entire universe depicted as a mechanical construction strictly obeying the mathematical rules of Newtonian mechanics, was the apotheosis of pragmatic science. But, as things turned out, it remained always just out of reach. From the very beginning, Newton's law of gravity had caused some philosophical disquiet, especially in France. The idea of a force of gravity was not in itself controversial. What caused unease, in Newton's formulation, was the idea of a force that acted, apparently instantaneously, across the vast emptiness separating one planetary body from another. What carried the force? By what means did the sun's gravitational pull reach out to the planets and exert a force on a distant body? On that score, Newton famously said *hypotheses non fingo,* I make no hypotheses. In other words, Newton was saying that his mathematical inverse square law of gravity worked very nicely, thank you, but why it did so was a question he forbore to answer.

French scientists of that era, as well as many others in Continental Europe, were very much under the influence of Descartes, who had proposed a qualitative model in which apparently empty space was filled with little vortices, tiny eddies and whirlpools, that carried physical influences from one place to another. The Cartesian model was never mathematical in the way that Newton's was. It merely proposed a picture to explain how not only gravity but also electric and magnetic forces, and light and heat, could travel across apparently empty space. The crucial point was that everything came down to local interactions, one vortex influencing its neighbors. Cause and effect thus traveled through a chain of connections.

Cartesians derided Newtonian gravity as "action at a distance," a pejorative term mocking the notion that one object could directly affect another with which it was not in direct contact. This, they thought, was philosophically objectionable, and they were not wrong. The necessity of action at a distance was a conceptual flaw. Newton didn't disagree, but simply passed over the point as a concern for future investigators. Meanwhile, his law of gravity, intellectually suspect though it may have been, made sense of the solar system for the first time.

Newton's chief cheerleader in France was Voltaire, who in 1738, half a century after the appearance of the *Principia,* wrote the *Éléments de la philosophie de Newton* with the deliberate goal of promoting pragmatic English science among Continental idealists. (Although the book has historically been credited to Voltaire, it owes a great deal to his companion Émilie du Châtelet, a remarkable philosopher and mathematician who was the first to translate Newton's *Principia* into French, adding her own commentary; that translation wasn't published until 1759, ten years after du Châtelet died while giving birth.) With a boost from Voltaire, Newtonian thinking made progress in France, and indeed it was French scientists who came to be associated most strongly with "action at a distance" laws for electric and magnetic forces. Coulomb's law, proposed in 1785 by Charles-Augustin de Coulomb, says that the force between two electrical charges varies with the inverse square of the distance between them, and is attractive between unlike charges and repulsive between like charges. The 1820 Biot-Savart law, due to Jean-Baptiste Biot and Félix Savart, says that the strength of the magnetic field generated by a current flowing in a wire is inversely proportional to the square of the distance from the wire. In both cases, a physical effect generated in one place manifests itself at another place in accord with a simple mathematical rule. In neither case is there any enlightenment as to the nature of that influence or how it got from one place to another. The Newtonian style of science had firmly embedded itself in the work of French mathematical physicists, with Cartesian concerns cast rudely aside once these savants realized how much progress a pragmatic cast of mind would yield.

A better understanding of the nature of electric and magnetic forces came toward the end of the nineteenth century. It is another tangled tale. In 1820, Hans Christian Ørsted, a Dane, discovered that an electric current would deflect an adjacent compass needle, implying that the current produced a magnetic force. Joseph Henry, in the United States, and Michael Faraday, in England, discovered the converse effect, that a magnet moving across a coil of wire generated an electric current. Clearly there were deep connections between electricity and magnetism.

In the 1820s, André-Marie Ampère made a first stab at a combined theory of electromagnetism, assuming the existence of a fundamental electrical particle; motion of those particles constituted an electric current and also generated magnetic forces. Ampère's theory was strictly of the Newtonian, action-at-a-distance variety, with influences apparently spreading instantaneously across space by some unknown mechanism. In an ironic twist, it was the Englishman Faraday who now rebelled against the Newtonian spirit of the lat-

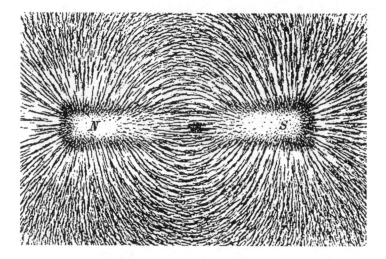

est Continental theorizing and offered a new way of thinking about electricity and magnetism—one that Descartes himself might have smiled upon.

Faraday was a rare type, the nonmathematical theorizer. The largely self-taught son of a blacksmith, he made his name with artful experiments that isolated numerous aspects of the interactions between electric and magnetic phenomena. As a result, he came up with a unifying picture—I mean literally a picture, mental imagery that expressed this thinking—to tie all those phenomena together.

Inspired, most likely, by the familiar pattern of iron filings around a magnet,[4] Faraday conjured up the notion of curving lines of forces that represented how the magnet's mysterious property of attraction and repulsion emanated into space. If a moving copper wire cut through those lines, Faraday proposed, an electric current would arise, in proportion to how closely packed the lines were, which was in turn in proportion to the strength of the magnet. The conclusions agreed nicely with Faraday's experiments. What's more, there was no action at a distance. The magnet's effect on the moving wire was mediated by the lines of force that the wire encountered directly.

He went further. He came up with the notion of an "electrotonic state" that pervaded all of space and which was the foundation of propagating electric and magnetic influences. Much scholarly ink has been expended on efforts to figure out exactly what Faraday meant by this. My own guess is that he was struggling to put into words an idea that made sense to him in a visual way but which he could not cast in quantitative terms. His electrotonic state was, I think, a medium in a state of physical tension, a sort of electromagnetic elasticity of space that tightened and relaxed depending on the currents and magnets in the vicinity.

This was not a theory, certainly not by the standards of the middle of the nineteenth century, when a theory was a structure of differential equations derived from Newtonian concepts of mass and motion. But what Faraday attempted turned out to be something profound. It was his entirely new conceptualization of electromagnetism, a physical model expressed in imagistic terms.

Most mathematical physicists of the day were baffled by Faraday's concept, even outright scornful. He was not one of their tribe. He could do arithmetic, to be sure, but he had no command of mathematics proper. Even today, he perplexes physicists who think of their subject in primarily mathematical terms. The Nobelist Emilio Segrè, in his popular history of twentieth-century physics, remarks in passing that although "Faraday wrote great physics without using a formal mathematical language, [he] thought mathematically."[5] Segrè doesn't explain exactly what he means by this, but I am pretty sure he is wrong. Faraday's imagination seems to have been primarily visual. By picturing the electrotonic state in his mind, he was able to further visualize how its influence pervaded space and what sort of experiments he might do to test his hypothesis. There was a quantitative element, to be sure. The electrotonic state could be tauter in one place, more lax in another, and that degree of tension must be linked to the magnitude of experimental phenomena he could look for. But when Segrè asserts that Faraday must have been thinking mathematically, he is testifying to the limits of his own imagination rather than the extensiveness of Faraday's.

Still, something mathematical had to be made of Faraday's conceptualization before it could be counted a real theory, and it was the mathematically sophisticated Scot James Clerk Maxwell who finished the job. Maxwell succeeded in transforming Faraday's provocative ideas into a coherent system of mathematical quantities and differential equations. One of the chief innovations of Maxwell's theory was the electromagnetic field, which was at least approximately a mathematical realization of Faraday's electrotonic state (not precisely, because Faraday imagined properties that the Maxwellian field didn't capture, and the field didn't exactly behave as Faraday had in mind).

Even better, as Maxwell explained, light itself turned out to be oscillations of the electromagnetic field. Another evident phenomenon, known since ancient times, had yielded to the power of mathematical physics.

It's no denigration of Faraday's genius to say that Maxwell was chiefly responsible for the classical electromagnetic theory in its rigorous form. But Maxwell himself, an occasionally waspish but fundamentally generous man, expressed his feelings clearly in a letter to Faraday: "As far as I know, you are the first person in whom the idea of bodies acting at a distance by throwing the surrounding medium into a state of constraint has arisen, as a principle actually to be believed in."[6] It was Faraday, in other words, who came up with the idea that electrical and magnetic influences are conveyed across space not by magic but by contortions and distortions of some kind of mediating entity—what we now think of as the electromagnetic field. And it's important to understand that the conceptualization came first, the mathematics later. Once again we see that mathematics is the language in which physical ideas are conveyed but not the *origin* of those ideas.

There's a curious but noteworthy backstory here. When Maxwell first attempted to turn Faraday's qualitative insights into a mathematical theory, he resorted to a highly mechanical model. He imagined space filled with little rotors (representing the twists of the magnetic field) that connected with one another by means of what he called idler wheels, sitting snugly between them. The spin of one rotor transferred itself to the idler wheel, which in turn caused the next rotor to spin. The function of the idlers was to ensure that the rotors all rotated in the same sense.

This sounds bizarre, but it was a standard method of theorizing: to construct a Newtonian model, you had to imagine a system of cogs and wheels, then figure out the laws it obeyed. It didn't mean that Maxwell truly imagined space to be filled with tiny machines in this way. Rather, it legitimized a theory by showing that it was founded on mechanical thinking.

In the end, though, Maxwell found that he could get rid of these gadgets and gizmos and replace them with a purely mathematical description of the electromagnetic field. The advantage was mathematical simplicity; the field was easy to describe. The disadvantage, at least to many scientists of that time, was that the field was a strange, newfangled beast, no longer visualizable in strict mechanical terms.

What, then, is this thing we call an electromagnetic field? It is defined, per Maxwell, as a mathematical entity obeying certain equations. Those equations dictate its behavior, its interaction with electric and magnetic objects, and so on. It was not a collection of particles pushed around by forces, nor was it an assembly of wheels and springs. A few scientists balked at Maxwell's theory for this very reason. The Irish-Scots physicist William Thomson had taken the first steps toward turning Faraday's thinking into a set of mathematical concepts and had prodded Maxwell to push further. Thomson later became Lord Kelvin, and as Kelvin he rejected Maxwell's electromagnetic theory and continued to tinker with his old-school ideas, envisaging electromagnetism as a system of moving parts, of wheels and springs and levers and cogs filling space.

Maxwell's theory triumphed, of course, and the notion of a field took root. As physicists became accustomed to dealing with the new theory, the electromagnetic field inevitably took on a certain kind of reality in their minds. They knew how it behaved, they understood what it did, and in time those mathematically fueled insights took on the nature of intuitions, of phenomena physicists could understand and predict in a qualitative way before doing the math needed to get the quantitative details right. (I like to imagine that Faraday came at it from the other direction, gaining an accurate but intuitive understanding of electromagnetism without the benefit of a mathematical theory.) In this way, the field attained a certain kind of understood reality, at least in the minds of physicists, even though they were no wiser as to its fundamental nature.

The story of electromagnetism is a contrast to the story of heat. Heat was at first mysterious, but was then revealed as nothing more than the motion of atoms—a perfect Newtonian denouement. Electromagnetism, on the other hand, began as a collection of strictly Newtonian forces acting on objects, and in the end was captured as a mysterious, pervasive field. Electromagnetic theory was logical and mathematical and rigorous, but it didn't fit comfortably into the universe-as-machine metaphor.

It is a widely held belief that by the end of the nineteenth century, physics was approaching what participants at the time thought was a state of near completion, if not perfection. Certainly, I acquired that belief somehow during my own education. Everything was understood, or nearly so; all phenomena in the physical world were explained, at least in principle (the ideas were right even if the math was too hard to solve). The idea of a mechanical universe, intricate and all-encompassing, reigned supreme.

Other sciences were getting in on the act. Chemistry was a matter of atoms and molecules reacting in predictable ways, releasing or absorbing energy in accord with the laws of thermodynamics. Geologists were becoming quantitative, measuring the properties of rocks and calculating how they might deform and transform on timescales measured in thousands or millions of years. Even Darwin's theory of evolution can be seen as an instance of Newtonian thinking, with creatures interacting in a complex system and the survival and extinction of species dependent on cause and effect at the scale of individual organisms.

In truth, though, the idea of a complete mechanical explanation of the natural world was never as compelling at the time as it has been made out to be in retrospect. The period of ascendancy, for one thing, was short. The principle of energy conservation was not fully understood until 1850; the second law of thermodynamics was nailed down only in 1865. Maxwellian electromagnetism dates to that same year. But even before the end of the century, in 1896, Henri Becquerel discovered radioactivity, an astonishing phenomenon entirely beyond the bounds of known physical theory.

What's more, as I have tried to explain in this chapter, the notion of a universe completely captured in a set of mathematical rules derived from mechanically based physical models was always a dream rather than a reality. The theory of heat showed how the model was supposed to work, a previously enigmatic substance satisfyingly explained as the Newtonian mechanics of molecules in motion. But the theory of electromagnetism had a huge unknown at its heart: the electromagnetic field was, in mathematical terms, plainly defined, but in physical terms utterly mysterious.

Physics, at the end of the nineteenth century, was like science at any time in its history, a mix of solidly established laws and deeply understood phenomena along with unexplained puzzles and arguable assumptions, all tied together with mathematical reasoning that was based sometimes on pure Newtonianism, sometimes not.

In any case, this state of affairs was short-lived.

PART III

FUNDAMENTAL PHYSICS CHARTS ITS OWN COURSE

The science of the nineteenth century was, by and large, the science of things we can see, hear, or touch, the science of the immediate and tangible world. The advent of quantum mechanics and relativity in the early twentieth century drew scientists into a world inaccessible to the senses, the world of atoms too tiny to see and energies beyond the capacity of laboratories to create. The content of scientific theorizing inevitably changed, but more important was a slow evolution, not always understood at the time, in how scientists thought about the world they were trying to explain and how they harnessed the power of mathematics to explore that world.

9

Dirac Invents Antimatter

I n 1931 the British physicist Paul Dirac wrote a scientific paper hypo-
thesizing the existence of a previously unsuspected subatomic particle. This doesn't sound, from our modern perspective, like such a
startling thing, but these were the early days of quantum and particle
physics. So unorthodox was Dirac's reasoning that he felt obliged
to begin his paper with a ruminative and somewhat philosophical
introduction justifying the argument he was about to make.

This was uncharacteristic of the man. Dirac was famously taci-
turn, and his scientific writings aimed for logical clarity, expressed
in the tersest terms. Of the rising generation of quantum theory pioneers, Dirac was the most purely mathematical. Graham Farmelo
titled his recent biography of Dirac *The Strangest Man,* and, although
I think that Isaac Newton surpasses him in that respect, anecdotes
testifying to his eccentricity abound. My favorite is the story about
him delivering a lecture somewhere and saying, as he chalked up yet
another mathematical expression, that what he had written was of
course obvious from what had gone before. An audience member
nervously spoke up: Excuse me, Professor Dirac, but is that last step
really obvious? Dirac stopped and stared at the blackboard. After a
few moments of awkward silence, he abruptly left the lecture room,
leaving the audience bemused. After a while—the interval stretches
to ten minutes in some versions of the anecdote—Dirac just as

abruptly returned, having given the question due thought. Yes, he confirmed, it's obvious, and proceeded with the lecture.

Dirac, that is to say, was not given to flattering his readers with needless explication, which is why the introduction to his 1931 paper stands out. Theoretical physics, he wrote, was embarking on a "process of increasing abstraction." Physicists could no longer put their trust in "direct attempts to formulate the experimental data in mathematical terms." Instead, he proposed, they would have to use "all the resources of pure mathematics in attempts to generalize and perfect the mathematical formalism that forms the existing basis of theoretical physics, and *after* [Dirac's emphasis] each success in this direction, to try to interpret the new mathematical features in terms of physical entities."[1]

To understand what he was driving at, and why he saw the need for change, we need to step back for a moment.

The discovery of radioactivity in 1896 proved to be the first of a series of upsets to the seemingly secure physical concepts of the late nineteenth century. Some wondered whether radioactivity, in which an apparently inert material spontaneously emitted strange forms of radiation, meant that the principle of energy conservation was being broken. Not until 1911 was it discovered that atoms, the still poorly understood constituents of all matter, contained tiny, heavy nuclei; and it was some years after that before radioactivity was seen to be an instability of certain nuclei that released energy trapped within. Understanding why some nuclei were unstable lay still further in the future.

Meanwhile, in 1900, Max Planck had come up with the provocative suggestion that the energy of light was divided up into little packets. Planck himself was doubtful that this was really the case, but in 1908 Albert Einstein offered additional arguments to say, yes, light came in the form of *quanta* of energy (which in the 1920s were christened "photons"). Quantum theory had arrived, and its emergence shifted the foundations of physics in wholly unexpected and disturbing ways. Quantum theory gave rise to Werner Heisenberg's uncertainty principle, which says we cannot even in principle find out everything we would like to know about the properties of an elementary particle such as the electron (which had been identified as a particle, a tiny object with mass and electric charge, by the British physicist J. J. Thomson in 1897).

Quantum mechanics in its full glory was invented in the mid-1920s, by Heisenberg and Erwin Schrödinger. In essence, it was an update of Newtonian mechanics for elementary particles, and it plainly stated that it was incorrect to think of such particles as little billiard balls knocked this way and that by conventional forces and collisions. Schrödinger's formulation of quantum mechanics described particles in terms of a wave function, a mathematical depiction amounting to a spread-out fogginess rather than a specific location. That fogginess was identified, a year or so later, as the probability of a particle's location at a certain point: the denser the fog, the more likely it was that an experiment would find the particle at that position.

Adding to the mystery, Heisenberg's version of quantum mechanics involved a novelty (to physicists, anyway) by the name of matrix mathematics. A matrix is a collection of numbers, such as a three-by-three array, and the rules for multiplying matrices together are different from the familiar rules of numerical arithmetic. In particular, the result depends on the order: when A and B are matrices, A times B is in general not the same at B times A.

It was quickly established, by Schrödinger himself, that his and Heisenberg's formulations of quantum mechanics were the same theory expressed in different mathematical language. Any quantity you can calculate in one you can also calculate in the other, although the calculation might be easier or harder. The novel physical concepts embodied in quantum mechanics, along with two different ways of doing the math, caused confusion that even now has not been altogether resolved.

This is not the place to get mired in the intricacies of quantum theory. The crucial point is that physicists had to contend with strange notions (a particle is not really a particle but a spread-out fog) and moreover had to learn novel mathematics to deal with these notions. Picturing the subatomic world in the classical mechanical sense, as a collection of intuitively familiar objects behaving in commonsense ways, wouldn't do. This was Dirac's point of departure in his 1931 paper. If you can't envisage the physics by means of appealing mental imagery, you have to lean all the more on mathematics.

Recall that the kinetic theory of heat, explaining that mysterious phenomenon as nothing more than the motion of atoms, began as a hypothesis about physics. Use of Newtonian mechanics transformed that hypothesis into a sound mathematical theory. The case of electromagnetism wasn't quite as straightforward. Maxwell began by imagining space filled with rotors and idler wheels that transmitted magnetic influences, and thereby developed a mathematical theory that also delivered a new concept, the electromagnetic field. The nature of the field wasn't so easy to grasp, but still, it was a definite thing that, with growing familiarity, came to seem like a respectable and visualizable part of the physical world.

With the advent of wave functions and matrix mechanics, on the other hand, commonsense visualization was of little help. To this day, if you ask a number of physicists to describe what they have in mind when they think about a quantum mechanical wave function, you will get a variety of answers. Some will use the imagery of a dispersed fuzziness, qualitative though it is. Others will insist that mathematics alone can tell you what a wave function does, and that to ask what a wave function *is* is to venture into useless metaphysics. The ineffability of the wave function is one of the things Dirac had in mind when he said that physics had embarked on a "process of increasing abstraction." He also said that physicists could no longer rely on "direct attempts to formulate the experimental data in mathematical terms." His point here is a little more abstruse. In quantum mechanics, the wave function is the central mathematical description of, say, an electron, but experiments do not measure the wave function directly. Instead, for a given experimental test, appropriate mathematical operations applied to the wave function yield the possible outcomes of that measurement and their relative probabilities. Turning the logic around, a collection of experimental outcomes does not, in general, point to a specific wave function; it can only ascribe probabilities to a variety of wave functions that are consistent with those outcomes.

A gap opened up, in other words, between measured quantities and the mathematical entities embodied in quantum theory. This is a dramatic change from the classical world, where measurements can be linked directly to mathematical entities, and where the more measurements you make, the closer you will get to a full mathematical picture. No matter how you visualize an electromagnetic field, for example, you can in principle measure it as fully as you care to and define it to any degree of mathematical precision that you choose.

But if, as with quantum wave functions, you can't visualize theoretical models in the old-fashioned, intuitive way, how do you proceed? Dirac's answer was that you must rely on mathematics to point you in the right direction. This, indeed, is how he arrived at his prediction of a new particle. A few years earlier, in 1928, Dirac had made his defining contribution to quantum physics: he devised an equation for the electron. It was remarkable for its combination of novel elements. It was a quantum mechanical equation, so that solutions to it were wave functions for an electron, describing how it would behave in given conditions, responding to given forces, and so on. Because the electron is tiny and can easily move at velocities approaching the speed of light, the equation had to incorporate aspects of relativity, one of Einstein's great achievements. Finally, it had to take account of a new quantum property known as spin, which like most things in the quantum world resembles its classical counterpart, but not exactly. The quantum spin of a particle has a direction, like the axis of a spinning top, but it also dictates how the particle behaves in a magnetic field (the notion of spin arose originally to explain why the characteristic wavelengths of light that an atom emitted changed a little in the presence of a magnetic field).

Dirac's electron equation was a strange beast, using novelties of mathematics previously unseen in physics to do its job. But it was spectacularly successful. Solving the equation allowed physicists to understand fine details of how electrons in atoms and elsewhere behaved. Dirac won a share of the 1933 Nobel Prize for his remarkable achievement.

But the equation came with a curious bonus. In any given situation, the equation coughed up two solutions. One, with negative electric charge, clearly corresponded to the electron that physicists knew. The other had a positive electric charge, and its spin was reversed, too. It would have been all too easy to dismiss this extra solution as an irrelevance, surplus to requirements. Such things were far from unknown in classical physics. The fact that the square root of a number has two values, one positive and one negative (the square root of 16 is either 4 or −4, for example) sometimes meant that a mathematical analysis of some complex mechanical setup might yield two possible solutions, but the traditional practice was to simply ignore the negative one as being clearly irrelevant to the problem at hand.

Dirac thought differently. He decided that if his ingenious equation was giving him two solutions, it was telling him something new about the world. He toyed with the idea that the second solution represented the proton, the positively charged nuclear constituent that was the only other elementary particle known for sure to exist at that time. But the proton is almost two thousand times heavier than the electron, and that was one of several reasons why Dirac's first idea didn't work. A physicist less invested in the power of mathematics might have let the question go, but Dirac was not so easily put off.

In 1931, he made the argument that the other solution to his equation was a real particle, but one that hadn't been found yet. He called it an antielectron, because that was exactly what it appeared to be—a particle identical to the electron but with its charge and spin reversed.

It's hard today to fully appreciate how daring Dirac's proposition was. From an empirical perspective, there was absolutely no need for the antielectron. Physicists at that time knew about the proton and the electron, and they were all but certain that there must also be something called the neutron, a particle similar in mass to the proton but without electric charge. Atomic nuclei were heavier than they ought to be for their known charge, if they were made up only of protons. The neutron was literally a makeweight to allow nuclei to have more mass than their electric charge would seem to indicate. There was no theory of the neutron (there was no fundamental theory of the proton either, for that matter), but experimental data pointed convincingly to its existence.

Dirac's case for the antielectron was just the opposite. It was demanded by no experimental evidence whatsoever, but mathematics, as far as Dirac was concerned, said it had to be there. His equation for the electron was the epitome of mathematical rigor, even elegance. It was impossible, according to Dirac's impeccable standards, that his equation could be lying to him. The antielectron simply had to exist, or the equation would be fatally flawed.

Even Dirac could see that this idea of making a purely *mathematical* case for a physical object no one had asked for, needed, or knew what to do with was a strange way of proceeding. This is why he started his paper with a philosophical preamble. He had to convince other physicists that his line of thinking was legitimate. As he put it, the theoretical physicist must use "all the resources of pure mathematics in attempts to generalize and perfect the mathematical formalism that forms the existing basis of theoretical physics, and *after* each success in this direction, to try to interpret the new mathematical features in terms of physical entities." His reason for italicizing *after*, I think, is that he believed the physicist should go to great lengths to demonstrate that a theory met all the required needs of the then-understood empirical situation. Only with that assurance was it then not only legitimate but necessary, in Dirac's view, to see what else the equations might have to offer. In the new world of quantum theory, a gulf was opening up between mathematics and experiment, and one must pay close attention to what the mathematics says, and conduct experiments to see if it is telling the truth.

Of course, I would not have recounted this story at such length had it turned out to be a wild-goose chase. In 1932, at the California Institute of Technology, Carl Anderson was searching among the tracks left by cosmic rays as they sped through a cloud chamber, a device filled with supersaturated vapor in which particles left a trail of visible droplets. He found a path that looked as if it had been made by an electron, except that it curved the wrong way in the cloud chamber's magnetic field—indicating a positive rather than a negative electric charge. Knowing nothing of Dirac's extravagant theoretical suggestion, he announced the discovery of what he called "easily deflectable positives," which in a fuller account he called "positrons," at the suggestion of an editor. Only after the result was published did others make the connection: Anderson's positron was Dirac's antielectron.

To put it shortly, a theoretical physicist predicted the existence of a new particle, and in due course an experimental physicist found it. We have become a little blasé about such tales, so often have they been repeated since then. But there's a first time for everything. What's momentous about the discovery of the antielectron was that it began with Dirac's insistence, on purely mathematical grounds, that the thing must be there. This was a real turnaround from classical physics, which always began with self-evident phenomena— heat and light and so on—and built mathematical theories to explain how they worked. Dirac did something new. So precious to him was the power of his mathematics that he determined the antielectron must exist, despite the absence of any empirical need. His insight turned out to have greater implications than even Dirac imagined at the time. Every particle, later theory made clear, must have an anti-particle. Matter has a whole mirror image counterpart that we call antimatter.

Dirac's invention of antimatter, at the time of its invention, explained no puzzling experimental results. It cleared up no confusions or discrepancies in the data. But it added to our understanding of the physical world, and it came out of mathematics, pure and simple.

10

Wigner's Enigmatic Question

D irac may not have started the modern revolution in particle physics, but he gave it an intellectual foundation. New subatomic particles arrived on the scene, singly at first but later in battalions. These newcomers had both experimental and theoretical origins. In 1930, the same year that Dirac predicted the antielectron, Wolfgang Pauli came up with another prediction, but for empirical reasons. In radioactive beta decays, the particles leaving the scene of the crime have less total energy than the particles that went in, and Pauli suggested that the missing energy was carried away by a neutral and elusive particle that was also produced in the decay. He called it the neutron, but in 1932, James Chadwick found the other neutron, the charge-free companion to the proton that gives atomic nuclei additional mass. Enrico Fermi renamed Pauli's particle the neutrino, the baby neutron, and although it wasn't discovered until 1956 it quickly became part of the furnishings of particle physics, so essential to theoretical coherence did it seem. In 1935, Hideki Yukawa, a Japanese theorist, proposed a new type of particle he called a meson, because it was intermediate in mass between the light electron and the much heavier proton and neutron. The justification for the meson was theoretical: Yukawa intended it as a sort of glue that kept atomic nuclei together. The positive charges of protons in a nucleus generate powerful electromagnetic repulsion trying to

tear the assembly apart, so there had to be a still stronger force—
the strong nuclear interaction, it was imaginatively named—holding
everything together. The following year, cosmic ray experiments (the
only way to study particles with very high energy, until accelera-
tors came along) revealed something that looked as if it might be
Yukawa's meson, but further investigation showed it to be a heavy
sibling of the electron—something no one had asked for, but there it
was. It was called the mu-meson at first, then simply the muon. Not
until 1947 was the first true meson found, the pion. Another meson,
the kaon, was found not long afterward.

That's enough of that. The emergence of particle physics in the
twentieth century is elegantly recounted by Robert Crease and
Charles Mann in their 1986 book, *The Second Creation*. The story
includes painstaking searches for elusive particles suspected to exist,
entirely unexpected discoveries, theoretical predictions that came
true, theoretical predictions that didn't, all in the context of attempts
to find order and organization among the proliferating parade of
mostly short-lived particles that were found as experimenters
learned to smash electrons and protons and atomic nuclei together
at ever higher speeds and sort through the resulting debris.

Even though physicists spoke of "particles" as if they were little
objects flying about, they had surely moved beyond the Newtonian
world of tiny billiard balls. To calculate how all these new particles
would behave, theorists had to describe them in terms of quantum
wave functions and calculate probabilities, not certainties, for this
or that outcome. The old word "particle" prevailed, out of habit and
convenience, but the underlying concepts were very different. A particle, in the modern sense, was not a little solid ball, but exactly what
it was was hard to say. Nevertheless, the mathematics of wave functions, as Dirac had foretold, proved a reliable guide, and as physicists
learned to handle the complexities of particle physics with increasing
ease, they were entitled to think that they understood what these
new particles did, if not precisely what they were.

Other mathematical innovations proved helpful, in particular the mathematics of symmetries and groups. To take a simple example, a cube has certain obvious symmetries. It can be rotated about its center through various combinations of right angles and will still look like the same cube. It can be reflected in a mirror, and you have no way of knowing which is the original cube and which is the reflection. The totality of all these symmetry operations, in mathematical lingo, constitutes a group. The essential notion is that any of the symmetry operations that belong to a group can be combined (performed one after the other, that is) to create a new symmetry operation that also belongs to the group. From a mathematical perspective, a group is a complete, consistent, and self-contained set of mathematical operations. For the theoretical physicist, it became a way of collecting related particles in a cogent way.

This part of the story begins with a scientific publication from 1931 by Eugene Wigner, one of several Hungarian Jews who eventually fled Europe for the United States. Wigner's argument delves into mathematics far hairier than I want to tackle in this book, but the gist of his achievement was to show how to turn arguments about symmetry into arguments about the properties of wave functions. That is, if a certain particle and its mirror image, for example, were known to be related in a certain way, then their wave functions must embody the same relationship. What made this result—it became known as Wigner's theorem—powerful was that it applied to any kind of symmetry, including not simply rotations and reflections in real space (the latter have the effect of reversing the sign of quantum mechanical spin), but also such things as changing positive charges into negative charges, or even reversing the direction of time.

Dirac's antielectron can be transformed into an electron by changing its electric charge and spin. They are not-quite twins. Wigner's theorem said their quantum mechanical description must likewise be closely related, so that if you reverse the direction of spin and the polarity of electric charge, the wave function for one turns into the wave function for the other. Soon after the 1932 discovery of the neutron, Heisenberg suggested that it and the proton have a somewhat similar relationship. One is charged and the other is not, and their masses are not quite the same (the neutron is a little more than a tenth of a percent heavier), but in other respects they are close siblings. Heisenberg proposed that they are different states of the same fundamental particle. Wigner coined the term "isospin" as a name for the relevant symmetry: the proton has isospin of +1/2, the neutron −1/2, so they are related by an isospin symmetry, or at least a near symmetry. Reverse the isospin, and one particle changes (almost) into the other.

The real power of Wigner's insight didn't dawn on particle physicists until a couple of decades later. By the 1950s, elementary (or seemingly elementary) particles were proliferating, and new "quantum numbers" were invented to describe them: charge, spin, parity (related to mirror image symmetry), isospin, strangeness, charm (the last a humorous characteristic bestowed on some particles that lived a charmed life, meaning that they lasted much longer than expected before decaying into other particles). It doesn't matter exactly what these quantities represent, except to say that they came out of efforts to classify particles by their properties, to organize them into kinship groups, so to speak, depending on how they interacted with each other.[1]

At first, these classifications were largely empirical, but as the number of particles continued to grow, it became an urgent matter to build a theoretical understanding of the relationships among particle properties. A number of physicists perceived that fundamental symmetries were at the bottom of the mystery, but finding mathematical schemes that lined up with the known properties of particles proved difficult. The symmetries involved were of an increasingly abstract nature. It's fairly straightforward to contemplate a mathematical operation that transforms a particle into its mirror image, but theorists now had to deal with operations that reversed a particle's isospin or its strangeness, when isospin and strangeness were enigmatic properties to begin with.

What first emerged, around 1960, was the idea that a particular mathematical symmetry group, going by the name SU(3), was at work. If you think of isospin and strangeness as numbers marked on lines at right angles to each other, then a particle with a certain combination of those two quantum numbers is a dot on the graph made by those two axes. The rules of the SU(3) group then tell you how to make transformations that move one dot to another's position. Each dot stands for a particle, and a mathematical group of transformations exist that turn any one dot into any other.

This method, associated most notably with the name of the American physicist Murray Gell-Mann, although many others played a part, led to the prediction of a particle that had yet to be found but was duly discovered later. The symmetry scheme insisted that a certain dot on the graph had to be there, otherwise the mathematical completeness of the idea would fail. The particle was found; the mathematics held true. This was Dirac's discovery all over again. If the mathematics is to work, if it is to retain its full mathematical coherence, a previously unsuspected particle must exist.

But there was a difference. Dirac had devised an equation, basing it on fairly well understood principles, to account for the behavior of the electron. That equation coughed up the antielectron. But the application of the mathematics of groups and symmetries in the 1950s and '60s was based on no fundamental principles, other than the fact—perfectly honorable, in the Galilean tradition—that it seemed to fit the available data quite nicely. The physicists of this era were somewhat in the position of Galileo who, after reasoning that a parabola was the correct shape for the path of a cannonball, lacked any true understanding of why the parabola was the shape that nature had selected. Newton answered that question with his inverse square law of gravity. In the 1960s the corresponding explanation, coming again from Gell-Mann, was the quark theory. Protons, neutrons, and mesons too numerous to mention were all, according to this proposition, different combinations of a small number of truly fundamental particles that Gell-Mann called quarks. The operations of SU(3) in relating all these composite particles to each other were then equivalent to switching quark identities in the various combinations. A proton is made of two up quarks and one down. Change one of the ups into a down and you have turned the proton into a neutron—giving substance to Heisenberg's original speculation that the neutron and proton were indeed close cousins.

That's all I want to say about quarks, except to add that they are never seen in isolation, and evidence for their existence, though compelling, is necessarily indirect—just as, in the late nineteenth century, evidence from the kinetic theory of heat for the existence of atoms was compelling but indirect.

Particle physics of the 1960s and later stretched theoretical physics beyond its Galilean and Newtonian roots, but not distressingly so. There was still the essential back and forth between theory and experiment: sometimes theorists predicted a particle and the experimenters found it—or didn't, in which case the theory was consigned to the trash heap of history. Sometimes experimenters found a new particle unexpectedly, requiring the theorists to come up with an explanation—except that sometimes the experimental evidence went away on further examination. There was the 1976 report of a new particle called the upsilon and later dubbed the "oops-Leon" in affectionate honor of its discover, Leon Lederman, the Nobel-winning physicist who had many real and lasting discoveries to his name.

But it couldn't help but be noticed that particle physics was moving beyond traditional physics in another sense: the use of strange new mathematics. The workhorse of Newtonian physics was the differential equation, which had quite literally been designed for the job. Newton invented calculus precisely because he needed a mathematical way to handle continuous change and the effect of force on motion. The utility of symmetry and group theory in physics in the twentieth century was not so easily accounted for. These devices recommended themselves initially as organizational tools capable of taming the chaos of elementary particles, but in the quark model, which posited the existence of new fundamental particles corresponding to the mathematical structure of the group SU(3), the mathematics of symmetry seemed to take on the role of a fundamental principle of nature. Many physicists learned to use these new devices from the mathematical toolbox without pondering the origin of their utility, but others, notably Wigner himself, wondered if deeper issues were at work.

In May 1959, Wigner delivered a lecture at New York University with the memorable title "The Unreasonable Effectiveness of Mathematics in the Natural Sciences."[2] The title summarized the question that nagged at him: How come so many tools of mathematics find their way into the physicist's description of the inanimate world? Wigner began by mentioning Galileo and his observation that two stones of any weight fall at the same rate. This was a regularity of nature, the stuff of which physical laws are made, and had in common with mathematical truths the virtue of being universal: it applied to all stones, dropped anywhere, at any time. Galileo's observation was a quantitative one, which naturally led to its being formulated in terms of simple mathematics, but in a case such as this, Wigner pointed out, "Mathematics . . . is not so much the master of the situation . . . it is merely serving as a tool." So far, so simple: the scientist notices regularities in the world and focuses on those that are amenable to quantitative analysis, involving properties such as mass and speed and time that can be measured numerically. It is no surprise that mathematical laws are useful.

But that's only the start of science, as we have seen. It took Newton's law of gravitation to explain why Galileo's observation was correct. That merely pushes the question one level deeper, however: How is it that the physical phenomenon of gravity could be encapsulated in a straightforward inverse square law, as Newton proposed? Remember that Newton himself was at pains to offer no suggestion for the origin of the inverse square law. *Hypotheses non fingo*, he said, and with that statement he separated himself from the philosophers of old, who would accept no proposition of a universal rule of nature *unless* they could come up with a fundamental justification for it, no matter how empty that justification might seem to us now (bodies fall, per Aristotle, because it behooves them to move toward the center of the universe).

We accept Newton's law because it works with such wonderful accuracy to explain falling bodies and the motion of the planets. But that's only half the story, Wigner observed. We also accept Newton's law because it is such a nice, simple piece of math, and because it is nice and simple we think we have discovered something basic and true about the universe. Does that mean, though, as Plato would have insisted, that simple mathematical laws are the only things the universe can be made of?

Wigner says that for the physicist, it is in essence an article of faith that physics can be couched in mathematical laws—it's pretty much a definition of physics that it is the search for such laws, and, so far, faith in that principle has been amply rewarded. But no amount of practical success can prove, to a logician's satisfaction anyway, that the principle is a priori correct. "The enormous usefulness of mathematics in the natural sciences is something bordering on the mysterious," Wigner says, "and there is no rational explanation for it."

Twentieth-century physics made the puzzle more puzzling still. As long as physics made use of mathematics that was reasonably close to empirical experience—numbers to measure things, calculus to analyze how things moved—it was possible to believe that science was merely picking out of the mathematical toolbox the tricks and devices most obviously suited to the task. But quantum mechanics and particle physics made that argument harder to accept. When he invented the matrix formulation of quantum mechanics, Werner

Heisenberg devised a kind of arithmetic in which A multiplied by B is in general not the same as B multiplied by A. That's alien to the arithmetic of the familiar world, and yet for quantum mechanics (which, after all, is not the mechanics of the familiar world) matrix arithmetic turned out to be just the ticket. Heisenberg found out only later that pure mathematicians had already invented matrix algebra, not because they saw any practical application for it but because it was where their imagination and invention took them. This, Wigner says, is the real puzzle. Mathematicians deploy wild inventiveness—Wigner goes so far as to call it recklessness—but they do so in a realm absolutely controlled by the iron laws of logic. You can be as imaginative a mathematician as you like, but what you invent has to play by the rules. In this sense, mathematics is an intellectual playground governed by its own strict laws but neither concerned with nor restricted by the happenstance and circumstance of the natural world around us.

And yet, as Wigner observed, mathematical tools invented in this way have a habit of proving useful anyway, and the more physics pushes into the subatomic world, the more arcane the mathematical tools it draws upon.

The burgeoning importance of symmetry and group theory (in which Wigner had a hand, although he didn't mention the subject at all in his lecture) was another example of pure mathematics elbowing its way into physics. Gell-Mann's pioneering efforts had something in common with Heisenberg's experience with matrix mechanics. In 1959, during a year in Paris, Gell-Mann hung around with mathematicians while he was struggling to find the appropriate mathematical representation of the apparent symmetries in the properties of known particles. It was hard work, slogging through all manner of different arrangements to find the one that would do the job, and he left Paris without success. Only when he was back in California did he find out that mathematicians had already done the hard labor for him. Despite talking to the right kind of mathematicians in Paris, and attending their lectures, he hadn't found what he needed. "The way they teach math is so abstract and peculiar," he said later, "it's very hard for a student to know what's going on. . . . They like to prove that there is something, but not actually *show* you what it is. When they give examples, they are so trivial that you don't learn anything from them."[3]

Scientists, on the other hand, need specifics. Generalities about the mathematics of symmetry are all very well, but if you are building a theory of known phenomena, you need the one particular example that does the job. Talking about mathematics as a toolbox may not be quite the right metaphor. Mathematicians write the laws and invent the concepts. More often than not, it's the physicist who has to use those laws and concepts to design the needed tool, and as we have seen with Heisenberg and Gell-Mann, they may well get most of the way toward inventing the tool without even knowing that the design specs for it are already tucked away in the mathematician's library, filed under some name that means nothing to the physicist.

In his lecture, Wigner did not arrive at any explanation for the unreasonable effectiveness of mathematics in the physical sciences. "Fundamentally, we do not know why our theories work so well," he said. "The miracle of the appropriateness of the language of mathematics for the formulation of the laws of physics is a wonderful gift which we neither understand nor deserve."

Almost thirty years ago, in *The End of Physics*, I pondered this question and came up with what I now think is an overly glib conclusion. I pointed out that physics, far more than other sciences, concerns itself with phenomena that can be quantitatively measured and said that physicists are "scavengers of mathematics" who use only what they need and pay no attention to the rest. I said that Wigner's puzzle reduces to a tautology: "Mathematics is the language of science because we reserve the name 'science' for anything that mathematics can handle."[4] I was pleased to discover that Bertrand Russell had said something along the same lines well before Wigner posed the question: "Physics is mathematical, not because we know so much about the physical world, but because we know so little: it is only its mathematical properties that we can discover."[5]

I am not so sure about this now. It's one thing to observe that science deals with quantitative regularities in natural phenomena, so that logic and mathematics are inevitably suited to their description. This is what Galileo thought, after all, when he said that mathematics is the language of the universe. But Wigner is asking a deeper question: Accepting that the natural world is described, scientifically, by quantitative and logical relationships, why should it be that what we prize as laws of nature are expressed in the form of appealing constructs from the world of pure mathematics?

There's an even bigger issue in play here. Einstein once famously remarked that "the eternal mystery of the world is its comprehensibility."[6] And Max Planck, before that, came to the same conclusion that Wigner reached: "Over the entrance to the gates of the temple of science are written the words: *Ye must have faith*."[7] Why, in other words, does science work at all? How is it that we are able to find quantitative, rational laws that mirror the workings of the universe? This is an old philosophical puzzle to which, so far as I know, there is no good answer. Like Wigner, we had better just accept that science works and be grateful.

Setting that aside, there's still the more specific question that Wigner asked about why innovations from pure mathematics find their way into physics. Implicit in that question, however, is the matter of how we choose what qualifies as a fundamental law. Suppose, for example, that Newton had worked out that the force of gravity must decrease not as the reciprocal of the square of the distance, but as some far more complicated formula, not easily expressed in basic terms. Suppose it had turned out to be some weird concoction of the Bessel functions I mentioned in chapter 7. Would we accept that concoction as a law of nature, or would we think that some deeper, simpler explanation must be hiding beneath it?

In his essay Wigner briefly refers to Einstein's opinion (though he gives no source) "that the only physical theories which we are willing to accept are the beautiful ones." If someone comes up with an accurate but ugly mathematical formulation of some problem, we may accept its utility but we resist thinking it is a fundamental law of nature. Wigner, however, is more interested in the fact that mathematical laws can be so accurate, and he turns away from the point that mathematical laws of nature ought to be pretty. The significance of beauty, however, has in recent times gained increasing attention in the community of theoretical physicists.

11

All This Useless Beauty

Theoretical physicists are apt to turn ruminative in their old age, and even Dirac, famous for his fierce devotion to rationality and impatience with wooly philosophizing, succumbed to the tendency. In a 1963 article on his views on physics that he wrote for *Scientific American,* Dirac remarked: "It is more important to have beauty in one's equations than to have them fit experiment."[1] He was not alone in this opinion. Hermann Weyl, a colleague of Einstein's who did much to advance the understanding of relativity, said, "My work always tried to unite the truth with the beautiful, but when I had to choose one or the other, I usually chose the beautiful."[2] Both Dirac and Weyl seem to be saying not merely that beauty is truth but that, ultimately, beauty is *preferable* to truth.

Measured against the perspective of classical science, these are strange sentiments indeed. Galileo would have been aghast. He had no patience with mystical blather about the divine elegance of mathematics, the sort of nonsense that Kepler embraced in his solar system of nested Platonic solids. Galileo spent his life putting mathematics to work in the service of facts and tried hard to extinguish the ancient philosophical idea that the power and rigor of math was the sole source of worthwhile truth.

In her 2018 book *Lost in Math,* the theorist Sabine Hossenfelder finds one case where physicists of the late nineteenth century enthused about the beauty of a theory. The theory in question was the vortex model of the atom, due in large part to Lord Kelvin and prompted by his seeing some clever experiments on the properties of smoke rings, in which they vibrated and gently bounced off one another. Kelvin proposed that atoms were similar formations in the ether, the hypothetical medium that was supposed at the time to carry light and other forms of electromagnetic radiation. Vortex atoms could oscillate in certain ways, explaining how atoms interacted with light; they could collide and bounce off one another, as required by the kinetic theory of heat; and they could link up to form molecules.[3] It was a highly appealing picture, "a theory about which one may almost dare to say it ought to be true," according to one contemporary, and one that "ought to be true even if it is not," according to another.[4]

The vortex theory was incorrect, of course, but in any case whatever beauty it might have possessed was in its physical conception, the way it tried to connect several distinct properties of atoms in a single model. Vortex atoms were an exercise in fluid mechanics of a very complicated variety, and the mathematics required to deal with them was distinctly gnarly. I was amused to find a comment on the subject by J. J. Thomson, the discoverer of the electron. For his doctoral work he had tangled with a problem in vortex atom theory. Like most such problems, he wryly remarked, it "involved long and complicated mathematical analysis, and took a long time."[5]

There was nothing attractive, in other words, about the mathematics of the vortex atom. In fact, from Galileo's time through to the end of the nineteenth century, I know of no physicist who rhapsodized about the beauty of mathematics in itself. Ludwig Boltzmann said exactly the opposite. "Elegance," he was fond of telling his students, "is for the tailor and the shoemaker."[6] Boltzmann used math to grind physical problems into submission, powering out the answer with any and all mathematical tools he could lay his hands on. What mattered was getting the right answer. Worrying about whether the methods or the results were beautiful would have struck him, perhaps, as prissy.

Pure mathematicians, who live in a world of their own imagination, have always had a thing for elegance. Bertrand Russell put it nicely: "Mathematics, rightly viewed, possesses not only truth, but supreme beauty—a beauty cold and austere, like that of sculpture, without appeal to any part of our weaker nature, without the gorgeous trappings of painting or music, yet sublimely pure, and capable of a stern perfection such as only the greatest art can show."[7] A soaring thought, to those who can appreciate it, but one with no evident relevance to the frequently messy business of ferreting out scientific truth. It's the old divide: mathematics is ethereal and self-contained, offering us a glimpse of heavenly perfection; science, though, must contend with the confusing and disorganized abundance of earthly phenomena, transient and seemingly haphazard, and latch on to small signs of regularity when it can.

Something happened in the transition from classical science to twentieth-century theoretical physics that caused some physicists to reach back (whether they realized it or not) to the ancient justifications for mathematical elegance as a criterion for knowledge, even truth. But before digging into that question, let us pause to see an example of the alleged beauty of mathematics. Dirac's surpassing achievement was his 1928 equation for the electron, an equation that combined the enigmatic quantum property called spin with the rules of relativity and thus dictated how an electron's wave function must be. Here, taken directly from his 1928 paper, is Dirac's electron equation:[8]

$$[p_0 + \rho_1 (\boldsymbol{\sigma}, \mathbf{p}) + \rho_3 mc] \, \psi = 0$$

You are forgiven if you fail to perceive the beauty in this formulation. To convey a sense of what this equation involves, you should know that the Greek letter ψ represents the electron wave function itself, the thing that can be interpreted, by the rules of quantum mechanics, to give the probability of finding an electron at this or that location, with this or that velocity. The parameters m and c are the mass of the electron and the speed of light. The symbol p is the electron's momentum and p_o is actually a differential operator that captures how the wave function changes with time—it represents, in a loose sense, the time component of the electron's momentum from the relativistic perspective in which time and space are regarded as partners in four-dimensional space-time. Lastly, the things designated by the Greek letters rho and sigma, ρ and σ, are matrices, or components of matrices, that relate to the electron's spin.

Don't worry if that makes little sense. The point about the equation above is that it is not, in fact, very elegant or attractive. It has a certain power, from the physics perspective, because of the way it puts wave function, relativity, and spin into one package, but the package itself is untidy. That's largely because it adheres to the classical tradition of treating space and time as unrelated. In his 1928 paper, Dirac spent several paragraphs proving that his new equation indeed obeyed relativistic principles. To do so, he cast the equation in a new form in which space and time were treated more equally. Here's the second version:

$$\left[i \sum \gamma_\mu p_\mu + mc \right] \psi = 0$$

A good deal neater, I hope you agree, but it's hard to see that this and the previous version are the same thing. The wave function ψ is still there, as are m and c, but apart from that it seems completely different. The trick is that the momentum of the electron is now treated as a single four-dimensional quantity, with elements p_1, p_2, p_3, and p_4, and the spin matrices have similarly been tidied up into a set of four matrices, γ_1, γ_2, γ_3, γ_4, denoted by the Greek letter gamma. In the second equation, the giant Greek sigma, Σ, indicates summation, specifically $\gamma_1 p_1 + \gamma_2 p_2 + \gamma_3 p_3 + \gamma_4 p_4$. The sigma notation saves ink and tedium when writing by hand or typing.

If you search the Internet today, you'll frequently find the Dirac equation expressed in still more terse formats:

$$i\hbar\gamma^\mu\partial_\mu\psi - mc\psi = 0$$

or even:

$$(i\partial\!\!\!/ - m)\,\psi = 0$$

These versions are more clearly related to the second one that Dirac wrote down. The differences are due to mathematical notations that put all the nitty-gritty detail into special symbols, rather than spelling them out explicitly. The final version is Richard Feynman's. He compressed all Dirac's cleverness with the relativistic spin matrices and differentiation with respect to time into a single symbol, the curly d with a slash through it, and called it the Dirac operator. Clever! And possibly beautiful!

I am not telling you all this in the hope that you will now fully understand the meaning of the Dirac equation. Rather, I present these different versions as a sort of aesthetic exercise, to illustrate something else that Dirac said: Beauty in math is something that "cannot be defined, any more than beauty in art can be defined, but which people who study mathematics usually have no difficulty appreciating."[9]

There's more to it than that, however. It may well seem that whatever beauty resides in the more compact version of Dirac's equation has been added through sleight of hand—redefining symbols and inventing new ones to sweep all the awkwardness under the carpet. To the mathematical physicist, though, these redefinitions make sense because they convey the neatness of a new formulation in which the symmetry of relativity (space and time given equal footing) and the innovation of the spin matrices (invented originally by Wolfgang Pauli but given an enhanced, four-dimensional relativistic form by Dirac) are explicitly built into the equation itself.

To put this another way, the Dirac equation, no matter how it is written, is not something that a pure mathematician would necessarily prize as an expression of delightful mathematical beauty. Whatever beauty it has comes from its ability to encapsulate *physical* meaning in a terse mathematical statement. It's different from the alleged beauty of the vortex atom theory, the mathematics of which was an impenetrable thicket. And it's different, too, from the kind of stark, austere beauty that Bertrand Russell was talking about. The mathematical physicist's conception of beauty does not necessarily line up with the opinion of the pure mathematician.

Dirac acknowledged that the beauty of mathematics is recognized only by those in the know, but suggests that all who know math will agree on what is beautiful. This is a questionable proposition. G. H. Hardy, an English mathematician of the first half of the twentieth century for whom the description "mandarin" might have been coined, wrote in 1940 a personal confession that he called *A Mathematician's Apology*. In it, he provides his own perspective on the power and beauty of mathematics, but a theme running through the book is his insistence that true mathematics, the best and most pure kind of mathematics, is marked by its uselessness. Pure math may find uses, of course, but that's by the way, and has nothing to do with its creation. Mathematics, for Hardy, must above all be *serious,* and applications of mathematics are to him decidedly unserious. Seriousness is allied with depth, which is allied with beauty, all of these characteristics forming the constitution of mathematics that deserves admiration and respect.

Of the tools that scientists make most use of, the differential and integral calculus, Hardy says, "These parts of mathematics are, on the whole, rather dull; they are just the parts which have least aesthetic value. The 'real' mathematics of the 'real' mathematicians, the mathematics of Fermat and Euler and Gauss and Abel and Riemann, is almost wholly 'useless.' "[10] He concedes that he is not totally averse to the idea that the theory of numbers, to take a standard example of pure and useless mathematics, might one day find some practical application. But he makes it all too clear that he finds this outcome highly unlikely, and emphasizes that the uselessness of number theory has been good because it allows mathematicians to stay above the fray, to set aside any concern that their ideas might be exploited by scientists and engineers. (To be fair, Hardy was writing this at the time of the Second World War, when scientists, engineers, and mathematicians were dragooned into military enterprises. His services, luckily for him, were never called upon.)

Hardy is greatly agitated by the question of uselessness. He says that "the great modern achievements of applied mathematics have been in relativity and quantum mechanics, and these subjects are, at present at any rate, almost as 'useless' as the theory of numbers. It is the dull and elementary parts of applied mathematics, as it is the dull and elementary parts of pure mathematics, that work for good or ill." In 1940, Hardy could just about get away with saying that relativity and quantum mechanics were mainly useless. Not so today. Electronics, lasers, computers, smartphones, and so on all depend on quantum mechanics. The GPS system that allows our smartphones to tell us where we are has to take note of relativistic aspects of the earth's gravitational field, so precise are its measurements.

Having noticed the intrusion of more advanced math into physics, Hardy continues: "No one foresaw the applications of matrices and groups and other purely mathematical theories to modern physics, and it may be that some of the 'highbrow' applied mathematics will become 'useful' in as unexpected a way, but the evidence so far points to the conclusion that, in one subject as in the other, it is what is commonplace and dull that counts for practical life."[11]

A Mathematician's Apology is a highly regarded book, but I find it in large part a self-conscious essay in academic snobbery, an effete paean to the superiority of intellectual work that takes pride in its inefficacy, as opposed to the toilsome slogging of those who are trying to understand the real world or even, Lord help us, the appalling efforts of engineers who set about making machines that do things.

I can't help thinking that what Hardy really meant was a reversal of Dirac's position: if any piece of mathematics turns out to have scientific value, then, alas, it can no longer be regarded as truly elegant. Hardy finds beauty in abstraction from the physical world, in uselessness; Dirac finds beauty (as did Wigner and Weyl) in the way math can make sense of the world around us, even give us a glimpse deep into the heart of physical reality. The enigma of beauty, in short, is not a matter on which mathematicians and physicists think alike.

Evaluations of the beauty of mathematics have, in any case, a mixed record within theoretical physics itself. A famous example is Albert Einstein's introduction of what became known as the cosmological constant. When, in 1917, he first used the equations of general relativity to model the universe as a whole, Einstein found that his theoretical universe insisted on either expanding or contracting. This seemed absurd to him, so he added a fix to his equation, in the form of a simple constant added to one side. This tweak made it possible to construct theoretical universes that were static and unchanging.

In 1929, however, Edwin Hubble came to the conclusion that the universe is, in fact, expanding. Observing many galaxies, he found that their light was redshifted—moved to longer wavelengths because they were speeding away from us—and that the magnitude of the redshift increased in an orderly manner with the galaxies' distances. The pattern, he concluded, was most straightforwardly interpreted as the result of wholesale cosmic expansion.

Now, of course, Einstein realized he had not needed to add the cosmological constant after all. Had he insisted, like Dirac, that the elegance of his original mathematics was beyond reproach, he could have predicted the expansion of the universe a decade ahead of Hubble's discovery. He later lamented, according to the folklore of physics, that the addition of the cosmological constant was the "biggest blunder" of his life. (It does not appear he said exactly this, according to a recent investigation, but he clearly expressed regret over disfiguring his equation in this way.)[12]

But the story changed yet again in the closing years of the twentieth century. Observers using the Hubble telescope found in 1998 that the recession velocities of galaxies at enormous distances, far greater than Hubble himself could have contemplated, were bigger than they ought to be. What this appeared to mean is that the universe is not expanding in a uniform way but that the expansion rate accelerates at the greatest distances. A simple way to account for this is to put a cosmological constant of the correctly chosen value back into the equations. (Einstein added the constant to prevent the universe from expanding, but depending on its magnitude it can also accelerate the universe's expansion.)

The modern interpretation of the cosmological constant is different from what Einstein originally had in mind. From his perspective, it implied that empty space under conditions of a perfect vacuum—no matter, no energy, no nothing—had some intrinsic curvature. In modern theories, on the other hand, the vacuum of space is filled with "dark energy," a strange kind of medium whose density (the amount of energy per unit volume) does not change even as space itself stretches out. Addition of a cosmological constant to Einstein's equation captures this property.

In short, what was once Einstein's biggest blunder, on account of its needlessness and ugliness, is now a standard element of cosmic theorizing, explicable in concrete terms as a new form of energy. What was once objectionable is now quite acceptable. The beauty or otherwise of the cosmological constant is a nonissue. It has practical value, and that's what matters.

The Nobel physicist Steven Weinberg has written that Dirac's admonition to search for beauty over empirical accuracy is helpful "only for physicists whose sense of mathematical beauty is so keen that they can rely on it to see the way ahead. There have not been many such physicists—perhaps only Dirac himself."[13]

But even Dirac was not infallible. In his 1931 paper arguing for the positive electron, Dirac offered a companion argument for the existence of something called the magnetic monopole. Whereas electric charge comes in two forms, positive and negative, that exist independently, magnetic north poles are always accompanied by south poles and vice versa. Isolated magnetic poles of either variety are unknown to science. Their absence leads to a certain asymmetry in Maxwell's electromagnetic equations. In the nineteenth century this was taken merely as a fact of nature, not a cause for lamentation, but Dirac, stirring some quantum mechanical ingredients into the mix, came to the conclusion that theoretical physics would be far more satisfying if magnetic monopoles, isolated north and south poles, were part of the zoo of elementary particles. They must have very high mass, he acknowledged, otherwise they would be created and seen in abundance. Since Dirac's time, particle physicists have come up with more sophisticated arguments for why magnetic monopoles would make the universe a better place. Experiments have sought to find evidence for such things on earth and in the heavens, but so far have found not the slightest sign of their presence.

So Dirac's prized sense of mathematical elegance led to the successful prediction of the positive electron but also the unfulfilled prediction of the magnetic monopole. It's worth quoting an apt remark of Niels Bohr, made in passing judgment on a very clever but ultimately unsuccessful theoretical proposal in the 1930s. "I cannot understand what it means to call a theory beautiful," he said, "if it is not true."[14] The saga of the cosmological constant perhaps illustrates a converse of Bohr's statement: if a piece of math turns out to be useful, perhaps it is beautiful after all.

Despite this murky and ambiguous history, the question of beauty in mathematical theorizing has grown into an increasing concern over recent decades. The reason, at bottom, is simple. Theoretical physicists have ventured into realms beyond the reach of experiment and observation. They propose hypotheses having to do with the elementary constitution of the physical world at scales so tiny and energies so immense that no terrestrial experiment can possibly hope to reach them. They search for theories that will combine quantum mechanics and Einstein's gravity in a seamless whole—the last remaining challenge of fundamental physical theory. They debate the nature of the big bang, a spectacular irruption in which our tangible universe somehow emerged from a fuzz of quantum uncertainty, perhaps creating space and time in the process.

In none of these ventures can direct experiment and observation be brought to bear. Theorists working in these overlapping realms, where quantum mechanics, particle physics, and general relativity all have a say, spend their time constructing ever more elaborate mathematical models. This is the world of theoretical exotica: strings, branes, hidden dimensions of space and time, holographic universes—recondite and mathematically complex ideas that may, we are told, hold the secret to a final "theory of everything" that leaves no part of the physical universe unaccounted for.

Whether you find these ideas enchanting, forbidding, or both, they are far removed from the kind of science that Galileo and Newton forged centuries ago, and far removed, too, from the simplistic "scientific method" that some of us learned as schoolchildren. Make an observation, formulate a hypothesis, devise an experiment to test the hypothesis. Repeat as necessary. In the modern world of fundamental physics, there is little if anything that counts as a novel observation, and nothing that qualifies as an experiment in the traditional sense. How do these theorists proceed? They rely on mathematical consistency and rigor, bolstered by a conviction that the correct theory must convey a sense of ineluctable rightness. It must be elegant, it must be beautiful—otherwise why would anyone consider it a final answer?

John Schwartz, one of the pioneers of string theory in its earliest incarnation, remembered of the early days that he and his collaborators "felt strongly that string theory was too beautiful a mathematical structure to be completely irrelevant to nature."[15] Admittedly, that's not as strong as saying that it *has* to be true, but it's a pretty clear statement nonetheless that nature only makes use of mathematically elegant theories, or even that any sufficiently elegant theory must be used by nature somewhere.

A large question looms ever larger: As physicists push on into a world beyond the reach of experiment, as they explore the foundations of the universe not with their eyes or their hands but with their intellects, as they rely increasingly on a sense of mathematical seriousness, or depth, or elegance, or beauty—call it what you will—are they, in fact, still doing science?

12

Science and Engineering

We hear a lot today about novel kinds of engineering. Genetic engineering, bioengineering, nanoengineering, even geo-engineering. These are cutting-edge activities, combining the latest scientific knowledge with exquisite experimental methods in projects that amount to rebuilding nature itself. Genetic engineers tailor the genomes of plants, animals, even human beings to create better (one hopes) versions of these organisms. Biological engineers, more broadly, build useful devices from biological ingredients, such as lab-grown skin for burn victims. Nanoengineers manipulate atoms and molecules to construct new materials and tiny sensors and machines. Geo-engineers, more controversially, propose global interventions to combat climate change, such as dumping iron into the oceans to boost the growth of plankton that eat up carbon dioxide (this probably doesn't qualify as an "exquisite" experimental method).

Whether these enterprises are good or bad I will leave aside. I have a more pedantic question in mind: Are they engineering or are they science? And what does that distinction mean?

When I was an undergraduate, we high-minded students of theoretical physics tended to look down on our engineering kin as the country cousins of real science—practical and ingenious but not quite as sharp or intellectually ambitious as our good selves. Scientists, we believed, and especially physicists, were engaged on the noble quest of discovering the secrets of nature. Engineers then took our knowledge and did things with it. It was also convenient to be able to blame engineers for any undesirable consequences of scientific innovation.

Truthfully, though, there can be no sharp distinction between science and engineering. Galileo himself employed his scientific knowledge to advise dukes and generals on matters of military engineering. The nineteenth-century inventors who steadily refined the technology of steam power happened on many scientific principles along the way. William Thomson, later Lord Kelvin, did much to establish the basic laws of thermodynamics but also helped design and lay the first transatlantic telegraph cable. The purely intellectual investigation of nature we call science, while the building of machinery we call engineering, but the one activity shades seamlessly into the other.

Nevertheless, the eagerness of many researchers today to openly call themselves engineers of one kind or another signals a change in emphasis, and one that has significant implications for the future of science. So far in this book I have told the story of the more obviously science-y aspects of science, from early astronomy and mechanics to thermodynamics and electromagnetic theory and on into quantum mechanics and particle physics. The distinguishing characteristic of these ventures has been, at bottom, the urge to understand nature. What we can call fundamental physics, involving the darkest corners of quantum theory, the beginning of the universe, and the search for the fundamental constituents of matter, represents the culmination of that long progress and the apotheosis of what we tend to think of as science in the purest sense.

But the vast majority of working scientists today care little for fundamental physics. All they need to know about physics is settled knowledge, secure and reliable. The idea of the atom as the elemental unit of matter is no longer strictly true, since we know that atoms are made of electrons and neutrons and protons, and we know further that neutrons and protons are made of quarks held together by glu-ons, and we suspect that quarks and gluons are not truly elementary either but built from even more arcane ingredients. . . . But really, for most scientists, none of that matters. If you are a materials scien-tist, for example, investigating the thermal and electromagnetic and optical and electronic properties of solids and liquids and whatnot, you certainly need to know a lot about electron physics, because the movement of electrons within a material is responsible for many of its characteristics. But you really don't need to know a whole lot about the nuclei of atoms beyond the fact that they have a certain mass and electric charge and spin. The fact that the neutrons and protons are made of quarks is neither here nor there.

The same is true if you are a chemist. Chemical reactions are in essence interactions between the electrons of one atom or molecule and another. The role of the nucleus is minimal. If you are a biochem-ist or a geologist, detailed knowledge of electron physics is probably not needed. You know the chemical properties of molecules, you know the physical characteristics of materials, and you don't need to know too much about how those properties arise.

For most scientists, in other words, atoms might as well be truly elementary objects. You need to know what they are and how they work, but you don't need to know the ins and outs of nuclear or particle physics, let alone string theory. There are, to be sure, nuclear physicists who try to understand how neutrons and protons arrange themselves inside a nucleus, and how the resulting properties of nuclei arise—these scientists need to understand quarks and gluons, but they don't, as a rule, dig much deeper than that. And the phys-ics of quarks and gluons inside atomic nuclei pretty much qualifies, these days, as settled science, regardless of our lack of knowledge about the ultimate constitution of those particles.

This may sound too pat, more complacent than science is sup-posed to be. In the twentieth century, the philosopher Karl Popper made respectable the notion that science is always contingent, always subject to change should new facts arise that unsettle what had previously been taken as settled. But Popper was writing in the aftermath of the huge upheaval caused by quantum theory, and he was concerned more than anything with what I call fundamental physics, the search for the deep laws of nature. Such upheavals are rare. In 1962 Thomas Kuhn published *The Structure of Scientific Revolutions*, in which he popularized the idea that there are two kinds of science, normal and revolutionary. Normal science connotes the refinement of existing theories, the improvement of experimental precision, the discovery and explanation of phenomena implied by standard theory but not yet demonstrated. Normal science is all about dotting the *i*'s and crossing the *t*'s, working out the next few decimal places. Revolutionary science, on the other hand, is when old laws of nature are thrown aside and replaced by new ones. The upending of classical mechanics by quantum mechanics is the prime example. There was also special relativity, which insisted that the speed of light is the same for all observers, and as a consequence revised our understanding of space and time. And then general relativity, Einstein's theory of gravity, gave space-time geometrical curvature, in contrast to the straightforward Euclidean space and immutable time of earlier scientific thought.

But Kuhn, like Popper, was sifting through the convulsions of early twentieth-century physics, a time in which the foundations of the natural world were rebuilt in a spectacular and thoroughgoing way. This era, however, was unique. The only comparable period is the foundation of science itself, when Galileo and Newton were establishing what we now know as science in place of earlier philosophical musing. The story of science between those two revolutions is, by and large, one of steady progress and accumulation. There were fits and starts, of course, and errors that had to be corrected, but on the whole, each succeeding generation of scientists built on the work of their forebears, and in turn set the stage for the next.

It seems to me that Popper and Kuhn were unduly influenced by their experience of living through the emergence of quantum theory and relativity. Kuhn's stark distinction between normal and revolutionary science doesn't survive close comparison with the progress of science since the time of Galileo and Newton. There is a spectrum. At one end, science is incremental, improving the measured accuracy of some fundamental constant of nature, devising a new wrinkle in the inordinately complicated theory of fluid mechanics, offering up an improved version of an old drug, and so on. Then there are bigger changes: recognition that a new value of some constant of nature requires a significant theoretical adjustment, the discovery of continental drift or of antibiotics, establishment of the structure of DNA and its crucial role in life. And bigger changes than that: the first statement of biological evolution, the suggestion that space-time has more than four dimensions. I could go on, and others might well put these and other advances at different places in the spectrum, from incremental advance to outright revolution. But there is no neat divide of the sort that Kuhn proposed.

Similarly, Popper's assertion that all science is eternally contingent, always vulnerable to change, is too sweeping. In principle he's right, but in practice, not really. Newtonian mechanics is still essentially what it was when Newton first wrote it down. The laws of thermodynamics are here for good. The basic principles of quantum mechanics have proved correct with such accuracy that they can't realistically change very much, no matter what the adventurers of string theory and multidimensional cosmologies find (this is analogous to the way that general relativity changed our understanding of gravity in a powerful way, but left Newton's law largely in place, sufficiently accurate for most practical purposes). The rapid ascent of molecular biology has seen significant changes in scientific understanding of the role and function of DNA, but it's unimaginable that there could be some other molecule, heretofore undiscovered, that's as central to life and evolution as DNA.

In other words, a great deal of scientific knowledge, once firmly established and verified by practical tests, is here for the duration. It may be subject to change, as Popper insisted, but only in tiny ways. There can be small refinements and tweaks, but the basic principles remain unchanged. (To be fair, Popper acknowledged that the more some part of science survived stern experimental tests and the more it proved its value, the more it became part of the canon, so to speak. But as far as I know he never allowed that any part of science could become unassailable.)

Wholesale upheavals in scientific understanding are today conceivable only in the realm of fundamental physics. Does the universe have more than four dimensions? If so, how many? Questions like this are genuinely unresolved, and the answers, if and when they come, will have enormous consequences. But questions of this magnitude are few and far between. The science of the world around us, as opposed to the sub-quark realm and far reaches of the cosmos, is settled science. It has been used so much, and with such wide success and accuracy, that significant change is inconceivable.

You may remember the saga of cold fusion when, in 1989, Stanley Pons and Martin Fleischmann claimed to have engineered a nuclear fusion reaction—the merging of two hydrogen nuclei to form a helium nucleus, the violently energetic reaction that powers the core of the sun—in a simple lab setting. Their claim brought much derision from the physics community, who said that if cold fusion was real, the laws of nuclear physics (and probably some of chemistry and materials science, too) would have to be rewritten. Well, that's too bad, said Pons and Fleischmann, suggesting, à la Popper, that theory was always at the mercy of experiment.

But it's not that simple. It would not have been a question of tweaking the rules of nuclear physics to allow cold fusion while leaving everything else intact. That would have been impossible. Cold fusion was so antithetical to established science that accepting it as true would have meant that astrophysicists could no longer believe in their understanding of the birth, life, and death of stars, and weapons designers would have to rethink whether they really knew why H-bombs exploded. In the end, cold fusion faded away because the evidence for it proved grossly inadequate, but physicists who declared, on theoretical grounds, that it could not possibly be true were by no means wrong to do so.

To come back to the question of distinguishing science from engineering, perhaps the best formulation is that science, traditionally, concerns itself with finding out the laws of nature, while engineering takes the truth of those laws for granted and uses them either to understand the world in more detail or to build new devices and machinery. If that's the case, though, most scientists today are not scientists after all but engineers. Few disciplines embody the nineteenth-century ideal of the natural world as machine better than modern molecular biology and medical science. Chemistry is a given. DNA is a given. The cellular structure of life is a given. Cancer is a malfunctioning of the human body, to be corrected by mechanical interventions—surgery, radiation, drugs. The drugs themselves are designed to derail the machinery of malignant cell reproduction while leaving the functioning of the rest of the body, as far as possible, alone.

To say that most of modern science is closer in spirit to engineering than to science in the Galilean sense is not to diminish its intellectual substance. The body of known science is huge, and even the best minds can master only a small portion of it. Saying that nature is a machine is far from saying that we know how the machine works—it is a complicated and intricate construction indeed.

Nor does the science-as-engineering perspective preclude the possibility of new phenomena. In 1986, for example, high-temperature superconductivity was discovered, in ceramic rather than metallic materials that were insulators in their normal state but abruptly turned into superconductors at temperatures well above what is possible according to the standard theory of superconductivity. Even today, scientists have not figured out exactly how these materials work. But no one expects a solution outside the bounds of conventional materials science and quantum mechanics. It's a matter of fully understanding the behavior of electrons in these exceedingly complex molecular structures—an engineering problem, in my terminology, like coming across a fantastically elaborate watch mechanism and teasing out how it works.

Even quantum computing, one of the most far-out research areas today, fits squarely within the concept of engineering. A quantum computer has bits that are not the stark zeroes or ones of a traditional computer, but "qubits" that can be a so-called superposition of zero and one, allowing such a device in effect to perform many calculations in parallel, and yielding a firm answer only at the end, when a suitable measurement of the quantum computer's state forces it to cough up a collection of definite bits rather than indeterminate qubits. As out of this world as this might sound, quantum computing strictly adheres to the rules of quantum mechanics, which belongs to the body of settled science, having proved itself to high accuracy many times over and for many decades. Building a quantum computer is a huge technical challenge, because it requires maintaining a host of interrelated qubits in a true quantum state, when all kinds of tiny disturbances and outside influences threaten to "collapse" the thing into a straightforward classical state. This is an engineering problem through and through, a matter of building a device that conforms to a well-known set of rules and does precisely the job you want it to.

Not every part of science, admittedly, fits into the science-versus-engineering perspective that I'm proposing here. Every year, to take a more or less random example, zoologists and botanists and entomologists discover thousands of previously unknown animal, plant, and insect species. This is science, and important science, too, as we contend with the effects of climate change and other human depredations on life on earth. Still, it's a kind of science that doesn't challenge fundamental assumptions about the physics and chemistry of life, and it sits easily with the concept of nature as machine, another example of the way that even in the twenty-first century we continue to discover new pieces of the mechanism.

A more consequential exception may be the vexed matter of artificial intelligence or, more broadly, the question of consciousness. Will machines ever be able to think in a way that mimics human thought, to the extent that they might pose as us or even surpass us? Can such a machine intelligence be conscious in the way that we are (passing over the fact that no one seems to be able to satisfactorily explain just what consciousness is and how you can test whether it is there or not)? Do new kinds of science await us in the exploration of these questions? Here I will simply nail my colors to the mast and declare that intelligence and consciousness are, fundamentally, matters of engineering, more precisely of scale. I mean that as computers get larger and more powerful they show signs of what can easily be taken as intelligence, and yet they are still a long way behind human brains in complexity and interconnectedness. Smartphones that answer our spoken questions aren't really smart, we might say, but they are pretty impressive, and very few users can clearly explain how they work. If you could go back even twenty years in time and demonstrate a modern smartphone to someone who had only just got used to the idea of small wireless devices that could make phone calls, that person might well believe the device was truly intelligent. Our standards evolve along with our machinery. I don't see any reason why a sufficiently powerful computing machine couldn't be regarded as intelligent and, if it was provided with sensors and inputs to know and respond to the world it lived in, conscious. That's my belief and I will leave it at that.

The broad distinction I am drawing between engineering and science harks back to the old philosophical terms "techne," meaning practical knowledge, the kind of learning showed by people who knew how to smelt ores into metal or make pottery; and "episteme," the sort of pure knowledge exemplified by geometers seeking to understand the structure of the heavens. In the ancient world, techne was the province of craftspeople and artisans, while episteme was the stuff of philosophy, and there was little connection between the two. Plato, with his idea of understanding the universe through mathematics and his positive disdain for practical knowledge, was the archetypal booster of episteme. Aristotle, who compiled enormous volumes on the variety of animal life, leaned more toward the techne end of the scale, although his larger purpose was the epistemic goal of understanding the true nature and purpose of life rather than such technical ones as improving animal husbandry or writing cookbooks.

Galileo's great achievement, in one sense, was to combine his appreciation for both episteme and techne and so create what we now call science, in which theoretical and practical knowledge are equally essential. Learning how to create and improve alloys, for example, has been the subject of a great deal of scientific research; nanoengineers today, among other things, are trying to custom-build new materials with desired properties by exploiting theoretical understanding with the practical ability to assemble structures by maneuvering selected atoms into place. Techne and episteme beautifully combined.

My contention, however, is that in recent decades the greatest portion of scientific activity has drifted closer to techne, in the sense that the theoretical foundations are secure and the crucial matter is putting them to use or understanding their implications in ever greater detail. What remains episteme in the modern scientific world is almost exclusively what I call fundamental physics, which continues the ancient quest to understand the nature of the universe in almost an abstract way. What Galileo joined together has begun to separate at the seams. Most of modern science is close in spirit to engineering. But fundamental physics has become a thing apart, a form, perhaps, of ancient philosophy in modern clothes.

PART IV

SCIENCE OR PHILOSOPHY?

If fundamental physics has detached itself, as I believe, from the mainstream of scientific work today and from the Galilean model that served so well for centuries, does it amount to science done in a new style, in an old style revived, or has it become a different activity altogether? To repeat the question I asked in the preface: What are scholars of fundamental physics today trying to achieve?

13

The Last Problems

In 1919, the Polish physicist Theodor Kaluza had a clever idea for making a unified theory of electromagnetism and gravity. He started with Einstein's general relativity, in which space-time is curved by the presence of mass: that curvature causes objects in free motion to travel along unstraight trajectories. We expect, following Newton, that objects should travel in straight lines. When they move along a curved path instead, we deduce that some force must be disturbing them, and in the case of a stone thrown into the air and falling back, we call the force gravity. What Einstein showed is that there is no force, but instead that curvature of space itself causes the stone to move along a parabola.

Kaluza added a fifth dimension, so that each point in Einstein's space-time (three dimensions of space and one of time) became a little circle. Think of that circle as a clock face with a single hand: the position of the hand could be connected, Kaluza showed, with the value of the electromagnetic field at that point. Instead of a theory of gravity with three dimensions of space and one of time along with a separate theory of electromagnetism, Kaluza had made a single $4 + 1$ dimensional theory in which both phenomena were intimately connected. It was, moreover, a fundamentally geometrical theory, building on Einstein's geometrical theory of gravity.

It's sometimes said that the history of physics is a story of ever-expanding unification, in which phenomena previously thought to be unrelated are found to be parts of a single system. Newton's theory of gravity unified the familiar terrestrial fact of apples falling from trees with the endless motion of the moon as it orbits around the earth. In thermodynamics, heat and mechanical work became different manifestations of an underlying property called energy. Maxwell put electricity and magnetism into a combined mathematical representation, and as a bonus found that light consisted of electromagnetic waves.

The notion that these advances were elements of a deliberate program by the name of unification is apparent only in retrospect, however. At the time, scientists were engaged in finding deeper connections between aspects of physics that were fairly obviously related. Once steam engines made their mark on society, it was hard to escape the conclusion that heat somehow engendered mechanical work; thermodynamics was the scientific explanation of that connection. Similarly, experimenters had known for centuries that electric and magnetic effects could influence each other; Maxwell's theory was the mathematical embodiment of those links. Putting such things together was the scientist's obvious task, a natural goal to aim for.

Declarations that unification per se is the basic task of fundamental physics began to appear only in the twentieth century. Perhaps Kaluza's attempted theory marks the onset of such ambitions, although his particular theory wasn't a success. For one thing, it was a purely classical theory, offering no concessions to quantum mechanics. In 1926, the Swedish mathematician Oskar Klein tried to remedy that failing by conceiving of the little circles as representing quanta of electric charge—a full circuit of the hidden dimension would correspond to the charge of an electron. But there was a big problem: since the Kaluza-Klein theory, as it became known, connected electromagnetism to gravity, this quantum of charge also had a mass, and that mass was enormous, something like the so-called Planck mass, about 0.02 milligrams, that results from a combination of basic physical units. That's bigger than the mass of the electron by a factor of 2 followed by 25 zeroes.

Although the original Kaluza-Klein theory quickly faded from view, one legacy of it is still with us—the possibility that there are more dimensions to space and time than meet our eyes. That idea lay dormant for several decades, however, as the proliferation of new particles starting in the 1930s took theorists in a different direction. Physicists discerned two new fundamental forces, the strong and weak nuclear interactions. The former is what holds atomic nuclei together; the latter has to do with beta radioactivity and neutrinos. Any particle can be characterized in part by its response, or lack of response, to these two forces and electromagnetism (the neutrino feels only the weak force, which is why it is so elusive). Unification now meant putting the weak and strong and electromagnetic forces into one tidy package. Gravity, of no measurable consequence for elementary particle interactions, was for the time being set aside.

A key element in this new attempt at unification was the introduction of broken symmetries. Symmetry itself helped make sense of the proliferation of particles by showing that they could be arranged in groups or families with related properties. Symmetry *breaking* proposed that, at very high energies, the three elementary forces are the same, and that differences emerge only as the typical energy of particles falls. The rationale for symmetry breaking is that it allowed physicists to make connections between elementary forces that are not obviously connected. The connection is broken in our low-energy world but restored, the argument goes, at very high energies. Unification of the forces in a single theory was by this point the overt goal of theorizing, and the concept of symmetry breaking became a deliberate part of the strategy.

The essential trick behind symmetry breaking is something you may have heard of: the Higgs boson, or rather, a whole bunch of

Higgs bosons suited for different purposes. The purpose of any Higgs boson is to break a symmetry. Modern quantum theory associates fundamental forces with particles. For electromagnetism, that particle is the photon, the quantum of the electromagnetic field. For the strong force, there is a set of particles dubbed gluons, which zip back and forth between quarks, binding them powerfully together. For the weak interaction, there are three such particles: the neutral Z and the charged W, which comes in positive and negative versions. Unification of the weak and electromagnetic forces was the first step. The idea was that in particle collisions at very high energies, the W and Z particles have no mass, just as the photon has no mass. Under these conditions, the weak and electromagnetic forces behave in very similar ways, and can easily be seen as different aspects of a single electroweak interaction. But at lower energies, the electroweak incarnation of the Higgs mechanism springs into action. The Higgs boson goes from being massless to acquiring a mass, and in the process it gives the W and Z particles masses, too. This is symmetry breaking—a state of affairs that was formerly equal, with photon, W, and Z all massless, changes into a new state where only the photon remains massless. It is the fact of the W and Z having mass that makes the weak interaction behave very differently from electromagnetism in the low-energy world we inhabit.

The W and Z particles had been found in 1983 at CERN, the European accelerator laboratory outside Geneva, and their measured characteristics fit nicely with expectations from the unified theory of the electroweak interaction. But as long as the Higgs boson itself remained undiscovered, a central question remained: Was there really such a thing as symmetry breaking, and even if there was, did the Higgs mechanism account for it or was some different theoretical cleverness required?

CERN's upgraded machine, the Large Hadron Collider (LHC), was built with the primary goal of finding the electroweak Higgs boson. Theory allowed the mass of the Higgs boson to be in a certain range, but as LHC experiments pushed to higher and higher energies yet still failed to find the particle, they pushed the Higgs mass close to the higher end of that allowed range. Finally, though, in 2012, the Higgs boson revealed itself, almost at the eleventh hour. Its estimated mass put it just within the range that theorists found acceptable.

There was a certain downside to this achievement, however. The Higgs mechanism is no one's idea of beautiful mathematics. It is ingenious, but it is a trick, a gadget, a kludge, as computer programmers say of a piece of code that is tacked on to a piece of software to perform some necessary but overlooked task. Explaining why the Higgs mechanism is unattractive would take us into technicalities. Suffice it to say that its only purpose is to break the symmetry inherent in a theory, and it does so by adding a new ingredient, the Higgs field, with properties devised entirely by the theorist in order to accomplish said symmetry breaking. There's nothing natural or inevitable about it, certainly nothing elegant. But it does its job.

In chapter 11 I gave examples of physicists pushing their theories into new territory by embracing a desire for mathematical beauty. The Higgs mechanism is a counterexample. It is not beautiful, but it is useful and versatile. Discovery of the Higgs boson of electroweak unification made it a respectable ingredient of particle physics more generally and bolstered the case for use of the Higgs mechanism in other aspects of the unification program. But its widespread adoption suggests that theorists' fondness for mathematical elegance is somewhat opportunistic: if an aesthetically unpleasing piece of theory does a valuable job and gains empirical support, then by all means use it; only when theory pushes into realms beyond the easy reach of experimentation does beauty take on a more significant role.

The next step in the unification program, at any rate, was to marry the electroweak force with the strong force and its gluons in what became known as grand unified theories (still excluding gravity, however). Versions of such theories exist, using a Higgs mechanism that operates at energies just short of a trillion times the collision energy of the LHC, and thus inconceivably far beyond terrestrial experimental capabilities. Particle physicists realized they would need to look elsewhere to test their theorizing, and so they became interested in big bang cosmology. During the earliest moments of the life of the universe, tiny fractions of a second after the big bang, the cosmos was a hot mess of particles at superhigh temperature. At the very beginning, if the grand unification program was correct, the electromagnetic, weak, and strong forces were all part of a symmetric scheme. As the cosmic temperature fell, first the strong interaction dropped out of the mix and went its separate way, at a cosmic age of about one-trillion-trillion-trillionth of a second, and later, at the advanced age of one-billionth of a second, the electroweak symmetry broke, leaving us in the realm of familiar physics. Physicists figured that these moments of symmetry breaking would leave their mark on cosmic history, and perhaps leave delicate clues that we might detect today.

Easier said than done, as it turned out. Cosmology could indeed separate the sheep from the goats, in distinguishing grand unified theories that were plausible from those that were not, but a whole lot of sheep remained. And then things got more complicated still.

In 1981, Alan Guth devised the "inflationary" universe, which in its first incarnation was a spin-off from grand unification. The Higgs mechanism works by causing a system of particles and forces to switch from a symmetric into a nonsymmetric state. This transition is triggered by a change in what physicists call the ground state of the system, in essence its lowest energy state, the one it occupies when nothing exciting is going on. Guth realized that this transition had cosmic implications, because the energy of the ground state mimics the behavior of the cosmological constant that Einstein introduced and later regretted. What's more, the change in that energy, when the grand unification symmetry breaks, is enormous. If, as Guth presumed, the cosmological constant today is low because the energy of the nonsymmetric Higgs ground state in which we live is low, then early on, before the grand unified theory Higgs mechanism swung into action, it must have been very large indeed.

The inflationary model of cosmology relies on these gravitational effects. The universe starts out in the high-energy symmetric ground state, and instead of transitioning smoothly to the low-energy state when it has cooled enough, it gets stuck in the high-energy state for a while. During this time the high value of the cosmological constant drives the cosmos to expand exponentially with time—hence "inflation." This inflationary period has, as Guth saw, two beneficial consequences: it explains why the universe today is so closely balanced at the critical point between expanding forever and eventually collapsing back in on itself, and it explains why the universe is, broadly speaking, so uniform in appearance (yes, there are galaxies and clusters of galaxies and vast empty spaces in between, but on the very largest scales the appearance of the heavens from any one vantage point is very much like its appearance from any other).

I don't want to get into the details of inflationary cosmology here. Guth's 1997 book, *The Inflationary Universe*, describes how it all works and digs into the tribulations that came after his initial realization. In particular, Guth's original hope that the grand unification Higgs mechanism would do the job proved infeasible, so physicists and cosmologists invented a wholly new Higgs mechanism specifically for the purpose of making cosmic inflation work as desired. Doing so was by no means easy, and this cosmic Higgs mechanism had to be designed with great care and scruple so as to produce a cosmology that fits what we know. This unfortunate necessity works against the original purpose of inflation. The idea was that a generic inflationary model would automatically and painlessly produce a universe looking like the one we inhabit, and thus do away with a lot of fine-tuning of the initial conditions at or soon after the big bang. Start with any old big bang you like, was the thinking, and inflationary magic would ensure that, ten billion years or so later, our universe is what you get. It has transpired, though, that making the universe come out right is a delicate matter, and all kinds of tinkering and fine-tuning have to be performed on the inflationary Higgs mechanism to make it so. This didn't eliminate the fine-tuning, it just shifted it from one place to another, and we will come back to this problem shortly.

The other big development was that as particle physicists discovered that cosmology offered them an arena in which to explore the feasibility and consequences of their theories, the drive to include gravity in the unification of the fundamental forces of nature gained new life. The big bang is in essence the moment when gravity became distinct from the other three forces. Before the big bang (we will skate over the question whether "before" has any meaning here), the density of the universe was so great that gravity was comparable in magnitude to the other forces. Once the big bang happens and cosmic evolution is on its way, the expansion of the universe follows the nonquantum rules of general relativity, while quantum theory must be used to describe the contents of the universe. This early period occupies "The First Three Minutes" of cosmological history that Steven Weinberg made famous with his 1977 book of that title. Weinberg brought the new science of "particle cosmology" to public attention. Once three minutes have gone by, the universe is made of familiar atoms and filled with electromagnetic radiation, and for the most part cosmologists and astrophysicists no longer need to pay attention to the exoticisms of unification theories.

Making quantum theories of gravity had been a sort of cottage industry going back to the Kaluza-Klein days. In the 1970s, for example, a theory called supergravity was proposed, the "super" coming from an earlier notion by the name of supersymmetry, which was in part an attempt to explain why the energies at which the strong force split off from the electroweak force and then the electroweak interaction split into its weak and electromagnetic components were so very far apart. In supersymmetry every elementary particle of the type known as bosons has to have a partner among the fermions.

(Bosons and fermions are distinguished by the amount of quantum spin they carry.) Among other things, supersymmetry created many opportunities for whimsical new particle names: the electron's partner was the selectron, the quarks were paired up with squarks, and so on. This was indeed a new kind of symmetry, and supersymmetric theories have a certain mathematical elegance—but to this day the direct experimental evidence for supersymmetry is exactly zero. A central theme of Sabine Hossenfelder's rueful *Lost in Math* is that supersymmetry has come to a tenuous position. There were many predictions, she says, that when CERN's LHC was up and running it would not only find the electroweak Higgs boson but also churn out evidence for the superpartners of known elementary particles. But as to that second ambition it has come up empty, and its failure has necessitated some theoretical rethinking. Supersymmetry is not ruled out, but the tweaking required to accommodate the LHC's failure to find anything super makes the theories more contrived and thus less attractive. The mathematician Ian Stewart remarked bluntly, "Supersymmetry is beautiful but it may not be true," and perhaps that will turn out to be right.[1]

Many theorists, even so, continue to believe that supersymmetry must be a fact of nature, and have taken it further. Supergravity is an extension of supersymmetry that includes the graviton—a hypothetical particle that bears the same relationship to the gravitational field as the photon does to the electromagnetic field. It has a superpartner, the gravitino. The first version of supergravity modestly required only four dimensions (three space dimensions plus time) to live in, but theories with additional hidden dimensions soon sprang up, though all had problems of various kinds (it was hard or impossible to make room for quarks; the cosmological constant in today's universe was horrifyingly big; etc.).

These theoretical innovations were mere preludes to what has become the dominant theme in fundamental physics today. In the so-called first superstring revolution, in 1984, theorists found that many of the difficulties of supergravity vanished if the basic objects in the theory were not particles but strings—literally, one-dimensional lines carrying a certain amount of energy per unit length. Only if these theories lived in ten dimensions did the superstring theory generate the array of particles and forces that we know, so six of the dimensions had to be wrapped up somehow to create tiny compact spaces, far below our ability to detect them, just as the old Kaluza-Klein theory of the 1920s made its single extra dimension into a tiny circle.

For a brief time it appeared that only one version of superstring theory met all the requirements for explaining our world, offering a possible answer to Einstein's probably apocryphal question: "What I wonder is whether God had any choice in the creation of the universe." But it quickly turned out that five superstring theories offered themselves up as equally plausible candidates. The second superstring revolution happened in the mid-1990s, when Princeton physicist Edward Witten found that the five ten-dimensional super-string theories could be seen as different manifestations of the same eleven-dimensional theory. So perhaps God had only one choice after all. . . .

But let's stop here. Accounts of superstring theories have been set out in a variety of books aimed at the general reader, the best of which to my mind is Brian Greene's *The Elegant Universe,* published in 1999. Note that adjective: the appeal of these theories has a great deal to do with their mathematical elegance. They seem, potentially, to have the ability to encapsulate everything we know of fundamental physics—gravity, electromagnetism, the weak and strong forces—in one perfectly consistent mathematically rigorous package. Greene's book is twenty years old, and string theorists have not been idle in the meantime. But it's also the case, at least in my jaundiced opinion, that the basic hope of superstring theory has not substantially advanced in that time, and that we are no closer to a true theory of everything than we were at the end of the twentieth century. As

Greene himself put it, in his 2011 book, *The Hidden Reality*, "As of today, then, the most promising positive experimental results would most likely not be able to definitively prove string theory right, while negative results would most likely not be able to prove string theory wrong."[2]

The titles of Greene's books offer a large clue to the conceptual difficulty with string theory and its offspring: he writes about an *elegant* universe, a *hidden* reality. The elegance of string theory manifests itself clearly only in its natural habitat, a high-energy universe of ten or eleven dimensions. What we can see, poor inhabitants that we are of a broken-down low-energy four dimensional world, are merely the ragged remnants of that intrinsic beauty, clumsy fragments of a perfect creation from which we hope to deduce, in circular fashion, the existence of an elegance that must always be hidden from our direct view. We are like Plato's cave dwellers of old, struggling to infer from the fleeting, cryptic shadows on the wall the hidden reality of a perfect, constant sun.

And it's not just symmetry breaking that creates problems. Perhaps the central conundrum of string theory is bundling up the extra dimensions of the ten- or eleven-dimensional universe to create the illusion that we live in a four-dimensional world. The difficulty is not in doing it; rather, it's that it can be done in so many ways, with no very good reason to choose one from another. Whatever splendid uniqueness string theory might possess in its pristine habitat is utterly lost by the time we come down to our universe. Brian Greene, in *The Hidden Reality*, says that the number of possible universes that can spring from a single version of string theory is around 10^{500}, ten with 500 zeroes after it. For comparison, our planet earth weighs somewhat less than 10^{28} grams; it contains about 10^{50} atoms; in the entire visible universe there are perhaps 10^{80} atoms. These numbers are unimaginably big, but they are still a very long way from 10^{500}. You could take every atom in the universe and imagine it was a little universe in its own right that contained another 10^{80} subatoms; then imagine that each of those atoms contained another 10^{80}; then imagine that each of those atoms contained another 10^{80}; then imagine that each of those atoms contained another 10^{80}; then imagine that each of those atoms contained another 10^{80}. You're still short of 10^{500} by a factor of 10^{20}. If you can wrap your head around that, you have a better imagination than I do.

Any fleeting hope that a string-based theory of everything would tell us how the universe must be or, in the words of Stephen Hawking, let us "know the mind of God" has long since departed.[3] Many physicists and cosmologists have given up that ambition and gone to the other extreme. Any possible universe that string theory permits must exist in some form, they declare, so that our universe is nothing special, just one out of 10^{500} universes or more that constitute the so-called multiverse. If that's the case, then many seemingly fundamental questions about the constitution and structure of our particular universe become empty. There's no point in asking why our universe is the way it is—it just happens to be that way, and there are other universes, beyond our ken, with different cosmic architecture. We will return to this issue in the final chapter.

Research into string theory and all its ramifications proceeds apace, but despite the enormous expenditure of brainpower over almost four decades, it's still far from assured that it holds the answers physicists and cosmologists are looking for. There is, to be perfectly clear, no such thing yet as a unified theory of quantum mechanics and gravity, containing all the elementary particles and forces we know and love. What string theory offers is the potential for such a theory to exist, assuming all the details can eventually be worked out in a satisfactory way. But there is no *proof* that such a theory exists.

Critics of string theory are in the minority among theorists who engage in fundamental physics, but a few are out there. Most notable is Lee Smolin, a former string theorist, whose 2006 book *The Trouble with Physics* offers an eloquent and historically astute analysis of how string theory arose and why, in his opinion, it gets too much attention. Although Smolin's book is more than a decade old, much of what it says remains true and valuable. Smolin provides a variety of technical criticisms and offers some insight into other possible avenues toward a truly unified theory; his 2001 book *Three Roads to Quantum Gravity* has more on those possibilities.

Smolin makes an acute point toward the end of *The Trouble with Physics*: string theory attracts the wrong kind of people. Since the early days of quantum mechanics, theoretical physicists have had to accept the intrusion into their discipline of increasingly recondite areas of mathematics. String theory takes that invasion to an extreme. The mathematical erudition required to master string theory and all its offshoots is forbidding to all but a few aspiring physicists. Moreover, it is essentially impossible to understand the principles of string theory *except* through the medium of mathematics. Here, for example, is what theorist Leonard Susskind said in an interview with the *Economist* in 2013 when he was asked how we can know that the extra dimensions of the theory exist: "It must be so because it is so deeply embedded in the mathematics and not easy to explain. I pride myself on being able to explain things to a broad audience, but sometimes you have to say it's buried in the mathematics. [The theory] just doesn't work unless you add six more dimensions to the world and we'll have to leave it at that."[4]

Contrast that attitude with the work of Michael Faraday, barely proficient in math, who came up with the profound theoretical concept of the electromagnetic field. Even James Clerk Maxwell, who made Faraday's idea into a true theory, began by picturing wheels and idlers carrying electromagnetic influences through space. Only when he felt he had a grasp of the essential *physics* of the electromagnetic field did Maxwell feel able to create the mathematical devices needed to portray it. In the early days of quantum mechanics, Werner Heisenberg reasoned his way from an understanding of electronic transitions in atoms to the need for a new kind of math, and only later learned that what he devised was the mathematics of matrices, unknown at that time except to certain pure mathematicians who were largely uninterested in physical questions.

Traditionally, the greatest theorists of bygone days have begun by reasoning about the conceptual physics of a problem, pondering mechanisms and ideas of cause and effect, and have then worked their way, often with great difficulty and many wrong turns, to a correct mathematical description. String theory, on the other hand, is in large part a matter of exploring and augmenting mathematical models to see what new wonders they might contain, and only afterward figuring out if those mathematical wonders could have any physical correspondence with the structure and contents of our universe. Even then, the wonders are far removed from our sight. This is Dirac's method taken to an extreme: push the math as far as you can, and see if you can shoehorn the phenomena of our world into it.

Smolin remarks that even Albert Einstein would be out of place in fundamental physics today. He was no great shakes as a mathematician. He cooked up the essential idea of general relativity by thinking as no one else had thought before about what it means to be weightless, to be in an elevator free-falling to the surface of the earth. If the elevator slowed down or sped up, a person inside would have the same sensation as someone in a train that was slowing down or speeding up. Gravity, he saw, is indistinguishable from acceleration, and that insight set him on the path to thinking about curved space-time. Only then did he turn to others to find out how curved geometry is understood by mathematicians.

String theory attracts what Smolin calls craftspeople, who are enormously adept at mathematics but lack the piercing vision of what he calls "seers"—scientists like Einstein or Faraday, or like Niels Bohr, who strove to make sense of the concepts of early quantum theory before digging into the technical implications. Seers are rare, of course, but Smolin worries that even such seers as might exist today would find the doorway to string theory closed to them.

String theory enthusiasts will no doubt object to this characterization. It remains the case, though, that while string theory has excelled in coming up with mathematical constructions that may or may not resemble various aspects of our physical universe, nothing that string theorists have done so far qualifies as a testable prediction in the old-fashioned scientific sense. The gist of string theory is that it is an attempt to come up with a satisfying mathematical account of things we already know, the word "satisfying" having no very clear meaning except that theorists will know it when they see it. Is this science as Galileo would have understood it, or is it an exercise in Platonic idealism? Plato sought mathematical perfection in the construction of the heavens and disdained the messy realities of the terrestrial sphere. In fundamental physics today, mathematical perfection exists only in the multidimensional universe, the modern equivalent of Plato's heavens. Our universe, the four-dimensional space we call home, now stands in for the irregularity and unpredictability of Plato's earthly domain, and the old problem crops up again: Does a mathematically beautiful theory of the ten-dimensional world have any useful connection to the phenomena of our universe, or is this mathematical understanding a self-contained goal, one that bears not at all on what we know of reality around us?

14

The Byte-Sized Universe

As far back as the middle of the nineteenth century, it became apparent that the new laws of the science of heat had implications for what we now call cosmology. This was at a time when the first law of thermodynamics, the conservation of energy, was by no means a settled principle, and the second law, the one about entropy always increasing, had not yet been formulated. Speculation about the universe itself, about the age and origin of the solar system, were matters more of theological interest than scientific. Even so, in 1852, William Thomson (later to become Lord Kelvin) published a short analysis with the title "On a Universal Tendency in Nature to the Dissipation of Mechanical Energy."[1] Drawing on Sadi Carnot's 1824 analysis of steam engines, which demonstrated that some heat is always wasted and rendered useless for the production of mechanical work, Thomson pondered the totality of all physical processes on the earth—animal, vegetable, and mineral—and concluded that the steady accumulation of waste heat must inevitably lead to a degraded state in which no further energy was available for conversion into mechanical effort of any kind. Our planet would become, Thomson wrote, "unfit for the habitation of man as at present constituted, unless operations have been, or are to be performed, which are impossible under the laws to which the known operations going on at present in the material world are subject."

Taken up and elaborated by others, Thomson's analysis led to the notion of the "heat death" of the universe, which says that the inexorable workings of thermodynamics in any finite system cause heat and energy to become spread out ever more evenly, ultimately reaching a state of perfect equilibrium in which no useful transfer of energy from one place to another is possible. There would be random jiggling around, as atoms crash into one another and exchange tiny amounts of energy, but nothing interesting could happen.

A few decades later, the same problem cropped up again in a new guise. Ludwig Boltzmann, the pioneer of the kinetic theory of gases, devised a statistical interpretation of entropy. In a volume of gas, the constituent atoms bounce around ceaselessly, so that an enumeration of their positions and velocities is an ever-changing catalog. The same outward state of the volume of gas (specified by volume, temperature, and pressure) corresponds to myriad different internal states. Boltzmann surmised that the more internal arrangements there are that correspond to the same outward state, the more likely that state is to occur. An unlikely state would be one in which all the sluggish atoms are at one end of a container, while all the speedy atoms are at the other—this would be a volume of gas that's cold on one side and hot on the other. The most likely state, by contrast, is what we call thermal equilibrium, in which all atoms everywhere are, on average, doing the same thing, so that the temperature is the same throughout. Thermodynamic equilibrium in this sense is the blandest, least interesting state the gas can be in.

Boltzmann's great achievement was to connect, via a precise equation, the number of ways the internal state of a gas can arrange itself with its overall entropy. Gas that's cold in one place and hot somewhere else is in an unlikely state; it has low entropy. As the atoms zoom around and mix up, the gas moves toward equilibrium; its entropy increases. Thermal equilibrium corresponds to maximum entropy, and is the inevitable end state of any finite system. The heat death of the universe is this basic idea writ large.

Boltzmann's new interpretation of entropy led to a somewhat testy correspondence with Max Planck and one of Planck's students, Ernst Zermelo. Planck, before he made his name by coming up with the idea of quanta, was a thermodynamicist in a very classical vein and uneasy with the intrusion of statistical thinking into physics. He objected to Boltzmann that if his interpretation of entropy was correct, then the existence of the universe as we know it—in a state very far from equilibrium, with stars churning out energy into empty space, and some of it being greedily taken up on the surfaces of planets to fuel life—seemed very unlikely indeed. The only solution, Planck said, must be that the universe began in an even more unlikely state, with lower entropy; and why should that be, if low entropy is intrinsically an unlikely circumstance? Boltzmann argued back, but his arguments, frankly, were not always clear or consistent. The problem lingered unsatisfactorily.[2]

As it lingers today, in fact. Immediately after the big bang, the universe was a seething cauldron of elementary particles, unimaginably hot and dense. An exotic state of affairs, to be sure, but at the same time a very simple state of thermal equilibrium—of maximum entropy. Today's universe is clearly not in thermal equilibrium, or it would be very boring and cold, with lethargic particles spread uniformly throughout. But does that mean that the evolution of the universe from big bang to now represents an overall reduction in entropy, a steady departure from thermal equilibrium toward disequilibrium? And what of Planck's assertion that the universe must have begun at a state of low entropy?

There's another strand to this argument, moreover, with even deeper roots. Isaac Newton saw that a collection of individual masses scattered throughout space was an unstable arrangement. If there happened to be a few more objects collected in a certain part of space, the excess gravitational force of that denser region would pull more objects toward it, increasing its gravity further and pulling in still more mass. It would be a runaway process, Newton understood, resulting in all the mass gathering in one place and leaving everywhere else empty—or, if the space were infinite, with isolated agglomerations of mass floating about far apart from one another.

Newton knew nothing of entropy, of course, but what he imagined seems like an outright denial of the second law. A uniform collection of masses in thermal equilibrium (we can imagine them moving about with a range of velocities) would curdle into a highly nonuniform distribution. I remember learning about this argument in undergraduate physics and being told that it illustrated a fundamental point: from a thermodynamic perspective, gravity can soak up a potentially infinite amount of entropy. Whatever reduction in entropy comes about because individual masses collect into a small number of large aggregations is exactly compensated for by an increase in the entropy of the gravitational field itself. This seemed like rather a pat conclusion to a puzzling business, but these were undergraduate lectures and we moved quickly on.

On top of that, there's the matter of the universe expanding. What does it mean to talk about the entropy of the universe, if the universe is infinite? The standard recourse is to talk about the observable universe, or the visible universe, or the universe within the cosmic horizon. These are different names for the same thing, and they refer to the volume of space contained within the distance light would have traveled if it set off immediately after the big bang. Moments after the big bang, the observable universe was very small, so that its total entropy (entropy per unit volume multiplied by the volume) need not be very large. As the universe grows and develops structure, the amount of entropy in a given volume of space may fall, but the overall volume grows. The second law says that the entropy within a closed system can only increase—but the observable universe is not a closed system, because it continually encroaches on adjacent regions of space-time. The underlying issue, once again, is how to fold the action of gravity (in this case, the cosmic expansion predicted by general relativity) into the calculation of entropy.

In his 2006 book *Programming the Universe,* MIT professor Seth Lloyd offers a more instructive way to look at this question, in terms of the growth of *information*—specifically, quantum information. In the 1980s, Lloyd had chosen for his PhD studies the exotic field of quantum information. So exotic was it that his advisers threatened to throw him out of grad school unless he could explain what he was up to. At the time, his fellow students of fundamental physics were all working on the latest craze, string theory, but as Lloyd put it, "I couldn't for the life of me see why what I was doing was any crazier than string theory."[3]

Information as a quantitative, mathematically well-defined concept came into being in the middle of the twentieth century. Claude Shannon, an engineer at Bell Labs, was studying the problem of transmitting useful messages—a phone call, say—along a line that was encumbered with electrical noise. In doing so, he defined rigorously what was meant by such information, in terms of a series of digital bits (0 or 1), and he thought about how to distinguish the wanted message from the noise, which could be represented as a series of random bits. Shannon's analysis yielded a formula for the information content of a message that was basically identical to Boltzmann's statistical formulation of entropy. Entropy was already understood to be a measure of the order or disorder inherent in a physical system. Shannon's groundbreaking analysis showed how the meaningful information in a string of digital bits, as opposed to the meaningless information in the form of noise, could also be regarded as a form of entropy. Broadly speaking, the meaningful information is comparable to energy that can be turned into useful work, while the noise represents the wasted heat, or unusable energy.

By the time Seth Lloyd arrived in graduate school, the quantitative treatment of information was well established in mathematical and engineering circles, but there was no suggestion that information theory had any relevance to fundamental physics. Lloyd realized, though, that for quantum systems information theory is a natural fit. Any such system lends itself to description in terms of bits—this part is in this state, that part is in that state, and so on—and a collection of bits can be straightforwardly interpreted as an amount of information. The additional wrinkle is that quantum states can exist in superpositions—this state and that state at the same time, as exemplified by Schrödinger's famous quantum cat, which is supposed to be both dead and alive until someone looks at it, forcing it to fall definitely into one state or the other.

Quantum computers rely on superposition. In a conventional computer, each bit is a definite thing, either a 0 or a 1. Quantum computers instead work with qubits, which are superpositions of 0 and 1, able to represent both at the same time. A single logical operation in a quantum computer is a highly controlled interaction of two input qubits that generates two output qubits. But since the input qubits are superpositions, the operation causes the 0 and 1 state of one qubit to interact with the 0 and 1 of the other—four operations at once, in other words.

Why does this matter? Scientists and engineers routinely use computers to simulate physical systems—that is, to perform calculations based on the laws of physics in order to predict the behavior of a given system. Computer simulations of the earth's climate and weather or the flow of air around a jet plane's wing are familiar examples. But all physical systems are, fundamentally, quantum systems. It's the interaction of atoms and molecules in a quantum way that drives all the macroscopic phenomena we see in the world around us.

A traditional computer can model the behavior of a quantum system, but only inefficiently. It needs to carry out independent calculations to keep track of all the 0s and 1s in true quantum states. But a quantum computer, Lloyd explains, can do the same thing *efficiently*. That is, a single qubit in a quantum computation can be made to correspond to a real quantum state in the system you are interested in. What's more, quantum computing is universal, meaning that any quantum computer with a certain number of bits can do the same jobs as any other quantum computer with the same number of bits—including the calculation of a physical system that is also defined by the same number of bits.

Now comes the payoff for this lengthy preamble. If a quantum computer can be put into a direct one-to-one correspondence with a given physical system, then the physical system is in a one-to-one correspondence with the computer. A physical interaction in the former corresponds exactly to a quantum logic operation in the latter, and vice versa. In short, any physical system *is* a quantum computer, constantly engaged in the process of calculating what it does next. The universe is a physical system. Therefore the universe is a quantum computer. The evolution of the universe is a vast quantum calculation.

To understand the implications of this conceit for cosmology, Lloyd introduces an additional fact: quantum processes *create* information. Think of a primordial quantum system, with a bunch of qubits in superposed states, all representing some mix of 0s and 1s. This is the state of maximum informational entropy, the most general arrangement possible, conveying the least information because it encompasses all possibilities. However, quantum uncertainty means that this unrefined mix would be a little hotter and denser in some places than others. And, because of our old friend gravity, the denser places draw other particles toward them, thus amplifying them.

Now the system begins to calculate. Qubits interact with one another and change their state. Some of them become 0s or 1s instead of indefinite superpositions. But these interactions proceed a little differently in denser spots than in sparser ones; indefinite states fall into definite ones in not quite the same way; structure develops; information grows. What this means, in macroscopic terms, is that the system overall gains specificity. It becomes informationally more interesting—it's a system in a particular state rather than one in a potentially universal but as yet undefined state. The totality of cosmic information grows. Lloyd acknowledges that we need a full understanding of quantum gravity to understand completely how this story begins and plays out, but the outline of an answer is there.

The full scope of this argument is more than I want to go into here. *Programming the Universe* tells the whole story. The upshot is that by seeing the universe as a quantum computer, it's possible to understand how an interesting structure, one with galaxies and stars and planets, with mineral and chemical processes on those planets—sometimes including organic processes, which may lead to the appearance of life and brains and so on—can arise from a seemingly information-free mess of hot particles. In fact, Lloyd argues, the emergence of information and complexity is not just possible but inevitable.

What I want to emphasize here is that *information* is at heart a purely mathematical concept. It is a way of describing the states of things—whether telephone messages, computer codes, or the universe itself—in a strictly numerical way, as collections of bits. Thinking of a physical system in these terms is equivalent to asking about the amount of information you would need to fully describe it or reproduce it. Information is thus a step removed from the classical physical description of systems as collections of objects interacting through forces. Information makes us think of the structure rather than the ingredients it is made of.

To be clear, Lloyd does not say that we should throw aside the old-fashioned physical way of describing things, as complex systems made of atoms and molecules, rather that the informational description is an enlightening adjunct to the old view. But a new question arises. If we think of the evolution of the universe and the emergence of structure as an exercise in the growth of information, are we seeing the reality of the cosmos and its physical content in a new and perhaps deeper way, or are we refracting the universe through our own current preoccupations? What I mean is that in the nineteenth century, at the time of the second industrial revolution, with technology and science growing hand in hand, it seemed altogether reasonable to think of the universe as a giant machine, an intricate mechanical construction, a factory of cosmic scale churning out myriad products. Now, in what has been called the third industrial revolution, driven by the advance of computation and information science, we are invited to think of the universe instead as a vast computer, an information-processing system of cosmic proportions. In particular, since the source of cosmic structure remains the same— gravity!—does the informational portrayal of the physical universe represent a deeper dive into physical reality, or is it merely another way to look at it, conditioned by our recent history and experiences here on planet earth?

The informational perspective, after all, does not make substantially different predictions about how the universe has evolved and will evolve. It offers a novel perspective on a picture that, broadly speaking, we already knew. To be sure, this is a problem with astronomy and cosmology in general. Astronomers study stars and galaxies in as many ways as they can devise, and come up with theories to explain their origin and properties. They can't experiment in the laboratory sense of the word, but their science is still science in a Galilean sense because there are lots of stars and galaxies to study. A theory or hypothesis makes sense only if it accounts for general properties of these objects—if it explains, say, why all stars burn, why some burn brighter than others, how their brightness relates to their size. The test of a theory is in how easily it accounts for the multitude of stars and their properties, with minimal recourse to tweaking of parameters and special pleading.

But we have only one universe—only one, that is, the protestations of multiversians notwithstanding, that we are aware of and can directly explore. The most we can ask of a theory of cosmogenesis is that it gives a plausible picture of how the universe we live in came to pass, consistent with what we have learned of the laws of physics from the world around us. How, then, are we to judge plausibility? To a nineteenth-century scientist or engineer, the universe as a machine was an entirely reasonable way of thinking. To us, the universe as a computer seems equally plausible and more up-to-date.

But the criteria for plausibility are our own, not the universe's. Are we seeing the universe for what it really is, or are we are making a metaphor that flatters our present-day sensibilities? That question hasn't received much attention—just as, in the nineteenth century, the depiction of the universe as a machine seemed so apt as to be hardly worth worrying about. But it's not just the universe that is subject to our repeated reimagining. One scientist, at least, believes that we need to rethink the very basis of fundamental science.

15

Is Math All There Is?

Fundamental physics is the preoccupation of a tiny fraction of all the scientists and engineers working today, but it captures a disproportionate amount of publicity because of both its exoticism and its standing as the culmination of the ancient search for the most basic secrets of the universe. It's no coincidence, I believe, that recent decades have seen the growth of a minigenre of books and essays about the power and beauty of mathematics. Petr Beckmann's *A History of Pi*, originally published in 1970 and still in print, offers a historical account of this number and its role in mathematics and science. The most successful book of this type is probably *The Golden Ratio*, a book from 2002 in which Mario Livio explores the history of a number that the ancient Greeks revered in geometry and architecture and which shows up in everything from number theory to natural patterns such as sunflower seed heads and spiral galaxies. Other books published since the early 1990s include *Zero: The Biography of a Dangerous Idea*, by Charles Seife; *e: The Story of a Number*, by Eli Maor; and *An Imaginary Tale: The Story of* $\sqrt{-1}$, by Paul J. Nahin. (*e* is the basis of exponential and logarithmic functions; the so-called imaginary number *i*, the square root of −1, shows up in everything from basic algebra to quantum wave functions.)

Parallel to these works are discussions of the intellectual appeal of math, often couched in terms of beauty or elegance. A 2002 collection of essays entitled *It Must Be Beautiful: Great Equations of Modern Science,* edited by Dirac's biographer Graham Farmelo, discusses why terse equations embodying deep ideas capture much of the spirit of physics today. More recently, Ian Stewart's *Why Beauty Is Truth* explores why aesthetic elements have played an essential role in the evolution of mathematical ideas, and Mario Livio's *Is God a Mathematician?* focuses on the question Wigner posed half a century ago: Why is math so darn useful?

In short, the spirit of Plato is abroad in the world again. Science, to be sure, still prizes practical accomplishment and useful invention, but as I suggested in chapter 12 these activities look increasingly like engineering and less like the basic science of Galileo and Newton and their inheritors. When it comes to fundamental physics and the science of the universe as a whole, mathematics has moved steadily to center stage.

No one has taken this idea further than Max Tegmark, an MIT professor who, as he admits, works on respectable questions of cosmology for his day job and indulges in deep thoughts during his off-hours. His book *Our Mathematical Universe* sets out to convince the reader not simply that the universe is governed by mathematics but that, in the final analysis, the universe is nothing but mathematics. What we perceive as physical reality is, according to Tegmark, pure mathematics alone. I will say upfront that I'm not convinced, but Tegmark's arguments are by no means outlandish—or at least, they start out reasonably enough but creep stealthily into strange and unfamiliar territory.

To begin, there is the old, vexatious, and still unsettled question of just what mathematics is. Livio, in *Is God a Mathematician?*, distinguishes three points of view. There is Platonism, which holds that the objects and propositions of mathematics exist quite independently of the human intellect, so that mathematicians do not create math but rather discover propositions that are already out there, so to speak. (The meaning of the word "exist" here is problematic, to say the least, and we will come back to it shortly.) Then there is formalism, which says that math is wholly an invention of the human mind, governed strictly by logic but proceeding in ways that are influenced by our preferred ways of thinking or, if you like, our intellectual biases. Livio mentions the example of his earlier subject, the golden ratio, which has an unambiguous mathematical definition but whose significance is by no means obvious: the ancient Greeks thought it was a big deal, but to early mathematicians in India it merited barely a footnote or two. Formalism declares that the ingredients of math are incontrovertible but that what we do with them is highly dependent on the human mind. (Ludwig Wittgenstein argued that much or all of philosophy can be thought of as a "language game"; formalism couches math in similar terms.) Finally, there is intuitionism, in which the basic elements of math have an empirical grounding. We see numbers of things and objects with shapes in the world around us; therefore we build math on a foundation of empirical arithmetic and geometry but take it to ever greater altitudes.

Livio makes a very reasonable case for a middle ground. There are parts of math that seem to have independent existence, in the Platonic style; there are parts that we invent, as formalism, through intellectual exploration; and there are elements of math influenced by the fact that our physical world is full of things that can be counted and measured. Tegmark, on the other hand, is a full-on Platonist. Livio cites Tegmark's description of a mathematical structure as "a set of abstract entities with relations between them" but then quotes a number of eminent mathematicians who clearly don't agree.[1] He concludes that Tegmark doesn't prove this fundamental point but simply assumes it.

According to Tegmark, "The fundamental Legos out of which everything is made appear to be purely mathematical in nature, having no properties except mathematical properties."[2] Later he says that "all physics theories that I've been taught have two components: mathematical equations and 'baggage'—words that explain how the equations are connected to what we observe and intuitively understand."[3]

To see what he means, let us go back for a moment to the nineteenth century and the mechanical metaphor for science, the picture of nature as a machine described by mathematical laws. Think of the common trope of the atom as a tiny billiard ball, careering around and smacking into other billiard balls according to the laws of Newtonian mechanics and, in the process, giving us the laws of thermodynamics. A billiard ball has mass, position, and velocity. To those qualities we attach numbers. If we want to be fancy we can think of the ball having color and density, perhaps even spin as well as linear velocity. Those qualities, too, can be designated scientifically in quantitative numerical terms. So the billiard ball has physical properties that are apparent to our senses, and for scientific purposes we describe those properties mathematically. Just like Galileo said: mathematics is the language of science. All well and good.

But now think of an electron. We may want to imagine it as a tiny object with mass and electric charge, but we know better—we have to think of it as, at best, a sort of fuzzy blob described by a quantum wave function and spread throughout space. But then how do we think of an electron's spin? Certainly it's not much like the spin of our familiar billiard ball.

Now think of a tiny loop of superstring, wriggling about in ten dimensions. Or think of the seventh of those ten dimensions. Do we have any real physical idea—that is to say, an old-fashioned, intuitive, mentally graspable idea—of what we mean? Hardly. And what of the elementary particles (not really elementary anymore) that arise from the different ways the string wriggles around? To the modern physicist, as Tegmark puts it, a particle is "an element of an irreducible representation of the symmetry group of the Lagrangian."[4] Not helpful to outsiders.

This is what Tegmark means when he says that the "fundamental Legos" of the physical world are mathematical: our physical concepts don't help us, and only mathematics can properly describe them. Hence, too, his description of scientific laws as mathematical equations plus baggage: the equation is the real thing, the baggage is metaphorical language we invent to make ourselves feel a little more comfortable with what the math represents.

But here again I think Tegmark is pushing the argument beyond the breaking point. First, there is not so clear a distinction as he imagines between mathematical properties and baggage. Recall that when Faraday came up with the idea of the electrotonic state to describe how electric and magnetic influences spread through space, he was not understood and often derided for burdening the precise world of physics with a vague metaphysical conceit. Even after Maxwell had made the electrotonic state respectable by transforming it into the electromagnetic field and finding equations that governed its behavior, there was opposition from traditionalists who found the notion of a field too abstract for their tastes.

It seems to me that the electromagnetic field is indeed not as physically real as mass and position and velocity—but there is nevertheless something real about it. Imagine taking hold of two bar magnets, one in each hand, and moving them toward each other. You will feel a force of attraction or repulsion that depends on how far the magnets are apart and how they are oriented. What you're feeling is not the field directly, of course, but the forces that the field dictates. Is the field a physically real thing or merely a mathematical construct? You tell me.

And what about entropy? It began as a puzzling quantity related to wasted heat that was not available for conversion into mechanical work in a steam engine. Then it became a property of the collective state of atoms constituting a gas, a mathematical quantity derived from their individual motions. Then it turned out to have something to do with information of all sorts. Is entropy physically a real thing or merely a mathematical construct? I don't know either.

The question of whether a physical quantity is real is decidedly a subjective one, dependent not only on our direct experience of the world but also on our *learned* experience, in the sense that electromechanical engineers who work every day with generators and dynamos and the like may well feel comfortable in describing the electromagnetic field as part of their lived reality.

What Tegmark seems to be saying, on the other hand, is that because we lack the intellectual and sensorial ability to describe physical reality at the level of superstrings and the like, that reality might as well not exist. This is a strangely anthropocentric point of view. It strikes me as a leap of faith, not an argument, to say that when it comes to fundamental physics, only the mathematics matters. It is, I think, a matter of perspective. Not all theoretical physicists think about theoretical physics in the same way. Some (Dirac is the prime example) apparently think in mathematics and try to figure out how the constructions they devise might be useful in describing the physical world. Others think primarily about the underlying physical concepts and look for the right mathematics to make their concepts quantitative and precise. I belong to the latter group, which I suppose explains my lasting admiration for Michael Faraday and his struggle to explain the electrotonic state to skeptical colleagues.

Here's a story that I hope will make this distinction clearer (people like me prefer stories to definitions). As an undergraduate, I was familiar with physical quantities that are scalars or vectors. A scalar is a simple magnitude, like mass or electric charge; it is defined by a single number, and that number is the same no matter how you look at the thing. A vector is a quantity, such as velocity, with both magnitude and direction. Although a given velocity vector is a fixed quantity, the numbers that define it depend on your frame of reference: you could say that a car is moving at 30 miles per hour to the east and simultaneously at 40 miles per hour to the north, or equivalently that it is moving at 50 miles per hour in a direction about 37 degrees east of north. Simple geometry relates the two pairs of numbers.

At some point in my education I was introduced to the next member of this mathematical family, which goes by the name *tensor*. I can't remember exactly how the introduction was made, but it was decidedly mathematical in tone. Our lecturer wrote down a 3 × 3 grid of numbers (a matrix), with each of its nine components representing a quantity relating to two of the three directions of space, denoted by x, y, and z axes; that is, the numbers were the xx, xy, xz, yx, yy, yz, and zx, zy, zz components of some quantity. Still with me? The lecturer then explained how, if we rotated our chosen axes to some new alignment, the numbers in the matrix would change to new values, although they still denoted the same overall *tensor* quantity. I was not the only student to find this mystifying. I remember that during one lecture a young man put up his hand and said plaintively, "But I still don't understand what a tensor *is*." To which the lecturer replied, with a touch of exasperation, "A tensor is an object that transforms like a tensor!" This was not the helpful answer he clearly hoped it would be.

Here is how I would explain a tensor. Think of a cube of jellylike material. You can bend it and twist it around in various ways that alter its shape but not its volume. You can hold the top and bottom of the cube and twist those faces relative to each other. You can slide the top face laterally with respect to the bottom. Or you can grab on to two corners of the cube, diagonally opposite to each other, and pull or push. It turns out that all distortions of the cube that leave its volume the same amount to some combination of these three moves, and that to describe all possible combinations you need a 3 × 3 array of numbers. This is the tensor that provides a complete mathematical account of the distortions. Typically, you would choose for the geometrical basis of the tensor the three axes of the cube, but there's nothing to stop you choosing other axes, and if you are interested in shapes other than cubes a different set of axes might be more appropriate. In other words, the actual physical distortion of a shape may be described by different matrices, depending on your choice of axes, but the numbers that make up the tensor are related by straightforward geometrical rules, just as was the case with the different ways of writing down the car's velocity.

In short, you need a tensor to describe the distortions of a three-dimensional object, and what makes this matrix a tensor rather than some random collection of nine numbers is that they change in a prescribed way when you look at them from a different geometrical perspective. A tensor, as our lecturer said, is an entity that transforms like a tensor!

For those who think in a primarily mathematical way, saying that a tensor is a certain set of numbers with certain properties under geometrical transformation is enough: it defines the thing precisely. But for those of us who have trouble thinking in such coldly logical and abstract terms, it's much easier to be given a particular example of a tensor and then have its mathematical properties laid out. From such an example, I and others who think in more visual terms are able to see how a tensor works in a general sense.

This is how I became comfortable with all the mathematics I learned as a student and in my brief career as a researcher. When it came to general relativity, for example, I conceived of curved space in a physical way, as an arena in which light rays followed curving paths and orbiting planets rolled around the inside of a bowl. From those undoubtedly simpleminded pictures I was able to understand the mathematical tools (tensor calculus, chiefly) by which Einstein was able to create a quantitative theory of gravity.

I say all this to make clear that I, unlike Max Tegmark, find it essentially impossible to think of physical theories and laws *only* in mathematical terms. I need the help of a physical picture to make sense of the math—and so I will continue to think of quantum mechanical particles as little fuzzy blobs and of superstrings as just that, tiny strings of energy. Tegmark's personal conviction that fundamental physics is most fundamentally captured by mathematical entities and structures alone is, to my mind, a diagnostic of his particular thought process, not a deep insight into the nature of physical reality.

Be that as it may, Tegmark comes to the conclusion "that if some mathematical equations completely describe both our external physical reality and a mathematical structure, then our external physical reality and the mathematical structure are one and the same, and then the Mathematical Universe Hypothesis is true: our external physical reality is a mathematical structure."[5]

To understand this perspective, we have to imagine a complete mathematical structure not simply as a collection of equations but as a specific landscape of numbers and quantitative relationships that extends throughout space and time. An elementary particle is then a line weaving its way through this landscape as it moves through both space and time. A human being is a collection of atoms, except that it constantly exchanges atoms with the outside world (breathing, eating, disposing of bodily waste, shedding skin cells . . .), so we have to think of a person as a sort of complex braided thread in the mathematical landscape, with fibers both entering and leaving it.

This is a heady picture, and it has a certain charm. *Our Mathematical Universe* sets the hypothesis out with great conviction, and delves into many more aspects and consequences of it than I want to describe here. My reluctance to accept the idea comes down to a specific issue: Tegmark never gets around to explaining what the existence of mathematical structures really means. Early on he declares that a mathematical structure is "a set of abstract entities with relations between them," which suggests that mathematical structures have an existence only within the confines of our minds. But by the end of the book he has turned that around to imply that our minds exist only as some sort of illusory phenomenon built from mathematical structures—which have now gained existence as the most fundamental reality. Very puzzling!

Think of a computer doing a calculation. It follows coded instructions, manipulating binary digits represented by tiny pulses of electric charge, and thus turns input into output. Computers from different manufacturers, with different hardware chips and different operating systems, won't do this calculation in exactly the same way. Those electrical pulses will rattle about following different patterns in different computer chips under different operating systems. Even so, they are all doing the same calculation.

But what, then, *is* the calculation? In one sense it is the specific set of electronic operations carried out by the computer, but those operations vary from one computer to another. In another sense it is more abstract, the collection of mathematical operations that it consists of. We use the two concepts more or less interchangeably, but they are not exactly the same.

A version of this ambiguity, it seems to me, bedevils Tegmark's mathematical universe hypothesis. We can think of a mathematical structure as an abstract set of operations corresponding to the cause and effect of some physical process. Or we can think of the physical process as the outward manifestation of a more basic mathematical structure. But it's not clear that "mathematical structure" means the same thing in both sentences, and to argue whether the mathematical structure or the physical process is more real leads to all sorts of unanswered questions about what we mean by existence and reality themselves.

The mathematical universe hypothesis is, I should emphasize, an extreme position. It has not been taken up by other physicists and indeed has attracted much criticism. But the fact that an accomplished scientist might think that his vision of a pure mathematical reality is a legitimate question for debate shows how far the modern obsession with math has gone, and how far away we have moved from classical notions of what constitutes science.

16

The Dream Universe

Scientists, on the whole, do not strategize far into the future. Nor should they. A biomedical researcher, to be sure, might dream of finding a way to block or impede a particular form of cancer, but a scientist who marches into the lab every morning and announces that, today, they will find the cure is almost certainly not someone to place bets on. The day's problem is likely to be much more about the nitty-gritty: Why does the protein created as a result of this specific genetic alteration affect a cell's function in so damaging a way? Why does my cell line keep fizzling out? Likewise, a nanoengineer may dream of building a larger-than-ever quantum computer, but the path to that success involves solving problems such as stabilizing a laser to make a delicately constructed coherent quantum state last a fraction of a second longer before falling apart.

The historical success of science comes from taking things piece by piece, one step at a time. Galileo struggled to figure out the path of a flying cannonball. Maxwell worked hard to tease out the precise nature of the connection between electricity and magnetism. Their achievements had enormous consequences, but that's not what drove Galileo and Maxwell and all the rest. They were up against pesky problems and they couldn't stand not knowing the answers. Thus does science inch painstakingly along.

Lately, though, as I hope the previous chapters have demonstrated, the arena I call fundamental physics has seen the entrance of grandiose ambitions. Or, more accurately, the revival of ancient but similarly expansive hopes. The philosopher's task, according to Plato and his disciples, was to use the power of reason to discern the very structure of the cosmos. Everyday phenomena were beneath them, too picayune for their consideration. Aristotle, contrary to accepted wisdom, had a more modern scientific instinct. He understood the necessity of gathering and collating information and making guesses about why things were so. But his guesses were mostly empty speculation. Despite this his name came to be associated with a rigid orthodoxy mostly not of his making. Plato, meanwhile, is seen by many fundamental physicists as the man with the plan, the thinker who sought out harmony and elegance and righteous truth as the means to cosmic enlightenment.

The nineteenth century had its vision of the universe as a grand machine, but scientists didn't expect that they would be able to draw up the full blueprint any time soon. It was God who created the machine, so most believed, and a scientist's job was to fathom one little part of it. It was the early twentieth century when grandiosity began to invade the thoughts of physicists. In chapter 9 I put the blame on Paul Dirac, who had the audacity to predict the existence of the antielectron from mathematics alone—and who turned out to be right. The change in the direction of physics was not his sole doing, of course, but he was an early harbinger of the trend. Later, long after his days of scientific glory were past, he began to wax lyrical about the glories and beauties of mathematics in the search for cosmic truth. And Dirac was far from the only one to succumb to delusions of grandeur. Albert Einstein was supposed to have wondered whether God had any choice in the creation of the universe—meaning, was everything about the cosmos dictated by basic laws. But that was the elderly Einstein, puffing on his pipe and trying to peer into the thoughts of *der Alte,* the Old One. Young Einstein, sequestered in the patent office in Bern, had his own pesky problem to solve: What would you see from the window of a train traveling close to the speed of light? That's where relativity came from, not from vague musing on the nature of the universe.

It was inevitable that once physicists began to contemplate the nature of matter and forces at an ever more distant remove from everyday experience, their methods would have to change. But having laid out in this book my perspective on how those methods have changed, I finally come to the big question: Are fundamental physicists still practicing science in an evolved but recognizable form, or have they slipped into another mode of thinking entirely? What are they trying to do, and can they ever do it?

In chapter 13 I explained why researchers in fundamental physics are striving to find a final theory, a seamless melding of quantum physics, particle physics, and gravitational physics. They have no choice: the lack of such a theory creates contradictions and inconsistencies that cannot be ignored. Early in the 1980s, when string theory burst upon the scene, there were hopes that it really would be a theory of everything—that the rules of mathematics would allow only one possible version of the theory with full internal consistency, and that everything about it would be fully determined, with no loose ends and no unknown parameters. As late as 2004, in his breathless book *Parallel Worlds*, string theorist Michio Kaku continued to express that hope. "My own view," he writes, "is that the verification of string theory might come from pure mathematics, rather than from experiment. . . . If we can finally solve the theory completely, we should be able to calculate the properties of ordinary objects, not just exotic ones found in outer space."[1] By way of example he says that if string theory allowed scientists to calculate from first principles the masses of protons and neutrons and electrons (and, one presumes, get the numbers right), we would be justified in saying that the theory had demonstrated its validity.

Such an outcome would make Plato very happy. But Kaku's hopes now seem absurdly optimistic. The most current view of string theory is that it represents an incomplete version of a still larger theory, imagined but not yet realized, that would have within it the possibility of many different kinds of physics. The prospect, in other words, that a theory of everything would uniquely predict our universe and everything in it no longer seems credible. That may sound disappointing, but it's what we should have expected all along. The ideal of a perfect and complete explanation of this sort is a distinctly prescientific notion, embodying Plato's belief that mathematics alone can fully determine the nature of the earth and the heavens.

Remember that when Kepler first began to think about the structure of the solar system, he imagined that the relative sizes of the planets' orbits corresponded to a nested set of geometric figures: octahedron inside icosahedron inside dodecahedron inside tetrahedron inside cube. This was Kepler's homage to Platonism. But as he studied observations of the planets, empirical knowledge refuted presumptuous idealism. Not only did the planets' distances fail to conform exactly to a geometrical succession, their orbits weren't even circles! Later, when Newton proved that elliptical orbits directly implied an inverse square law of gravity, it would never have crossed his mind that his new law had the capacity to specify the sizes of our planets' orbits. All ellipses are possible. The orbits we know are a small number out of that infinite set of possibilities. Why the planets should have the particular orbits we see was, from a nineteenth-century or earlier perspective, an act of God, or, from a modern perspective, mere happenstance resulting from the complex process by which a spinning cloud of gas and dust condensed, about four and a half billion years ago, into our solar system.

This is exactly how modern science broke free of the hold of ancient philosophy. Scientific laws, deduced from observation and experiment and worked out in the language of mathematics, tell us how things work, how phenomena happen. They don't generally tell us how things come to be. Within any given discipline, certain facts and information are taken for granted. Oceanographers try to figure out how ocean currents flow, energized by solar heating and influenced by the earth's rotation, but they don't think about why the oceans have the shapes and sizes they do. Geologists studying continental drift try to understand how the oceans change over time, but they don't give too much thought to the formation of the earth. Planetary astronomers want to know how our system formed, but they take the presence of interstellar gas and dust clouds as given. And so on. The vast majority of working scientists stay well away from the issues that concern the theorists of fundamental physics and cosmology.

In the late nineteenth century, when a few scientists began to contemplate the origins of the cosmos, they had an easy out. The Creator made the universe and the laws that govern it, and it was the task of scientists to find out those laws, the better to worship him. Einstein's God was not a Creator in the old-fashioned sense. It was Einstein's name for a guiding principle that ran the whole show. But if, as Einstein hoped, the laws of fundamental physics allowed one universe and one universe only to come into being, another question pops up: Why should that universe be so finely attuned to our existence?

Martin Rees's 2000 book *Just Six Numbers* shows how remarkable that fine-tuning is. Rees describes how certain fundamental parameters, such as the overall density of matter in the universe and the relative strength of the electromagnetic and gravitational forces, play a surpassingly important role in determining what kind of structures (from galaxies to planets) are possible, what array of chemical elements the universe will create, and whether life is possible or not. These numbers—which are parameters of physics, not ingredients of pure mathematics—cannot stray very far from their known values, or the universe as we know it would not have come about. It would be devoid of stars that lived long enough for life to happen; or it would be filled with hydrogen and hardly anything else; or such planets as exist would be tiny, rocky, and barren; or chemistry would be inadequate for the complexity of life as we know it. Etcetera.

There's a glib response to this, which is that of course the universe is congenial to life, because if it weren't we wouldn't be here to fret about it. That's true in a punctilious sense but it hardly amounts to a satisfying answer. If it's true that the laws of physics are so carefully adjusted as to make the universe and life within it possible, and if there's only one universe made by those laws, then we are right to be suspicious. A religious believer can say that this is evidence of God's hand, but that won't fly for most scientists these days, and in any case, even if you believe it, it's not a *scientific* answer.

Almost inevitably we seem to be driven to the multiverse hypothesis, which says that our universe is one of an almost inconceivably large number of universes, and that the physics within our universe is merely a single example of the kind of physics that the overarching theory can produce. Other universes would have different elementary particles, different forces, and, as a consequence, different versions of basic physics and chemistry. They might have planets or they might not; those planets might harbor life, or they might not. Rees embraces the multiverse idea, as does Tegmark, who offers an appealing general justification that he heard from Alan Guth.[2] Guth made the point that as soon as we start thinking about the creation of the universe as the result of a physical process, it's unavoidable that such a process would have the capacity to give rise to various universes with different properties and characteristics. This is no more than the reasoning behind Newton's explanation of the planets' orbits: gravity allows all kinds of solar systems to exist, and ours is merely one of the possibilities. Likewise, a set of laws governing how universes arise will, if it's consistent with how other physical laws work, allow many kinds of universes, and ours is merely one of the possibilities.

But there are pitfalls in taking these arguments to the cosmic scale. Many astronomers and cosmologists embrace what they call the Copernican principle. Copernicus demoted the earth from its prior position at the center of the universe and put the sun there instead. The larger lesson is that our planet and our species are not at the center of all creation, and in due course this extended to the realization that our sun and our solar system are also nothing special. Likewise, we live on the outskirts of an ordinary galaxy, one of many billions looking much the same. The Copernican principle—which Copernicus himself, of course, never enunciated—is that we should be wary of any theory that requires us to occupy some sort of special, privileged place in the universe. Going further, some cosmologists suggest, we ought to believe, too, that our universe itself is merely an ordinary, run-of-the-mill inhabitant of the multiverse, marked out in no special way.

But as Rees says, "We shouldn't take Copernican modesty (sometimes called the 'principle of mediocrity') too far. Creatures like us require special conditions to have evolved, so our perspective is bound to be in some sense atypical."[3] In other words, our universe really is special insofar as we are in it. This is not a statement of self-aggrandizement or cosmic speciesism, merely a recognition that the conditions that allow life to emerge are by no means assured in every universe within the multiverse, especially when we allow that physics itself could be very different in other places.

Some enthusiasts for the multiverse like to argue nevertheless that the physical parameters of our universe are, in a broad way, likely or at least relatively common as compared to the corresponding parameters of other, non-life-friendly universes. Steven Weinberg, for example, has argued that the value of the cosmological constant indicated by dark energy measurements may not be as arbitrary as it appears, because if it did not have a value in that ballpark the universe would be a much less likely habitat for life such as ourselves.[4]

The point, allegedly, is that our universe is not so special and by extension that the multiverse hypothesis itself is not such a stretch. But arguments of this sort strike me as fatuous. Think of that number 10^{500}, the number of possible universes within the multiverse, and suppose that universes broadly like ours are reasonably common— one in a million, let's say. If that's so, the number of universes that could be our home is 10^{494}—a very big number! Now suppose instead that universes like ours are exceedingly rare, so that only one in 10^{400} can house us. Even then, there would still be 10^{100} universes in the multiverse where we could take up residence. That's still a very big number, many times greater than the number of atoms in our universe. Does the first scenario seem more reassuring than the second? Is there any way we could tell the difference? Do these numbers matter at all?

The overwhelming difficulty with the multiverse hypothesis is that if there is only one universe that we can know about, how can it be a legitimate scientific proposal to say that there are innumerable other universes out there, in some space that we can never explore, the existence of which we can never definitively establish? A legitimate scientific theory, we learned from Karl Popper, is one that can be falsified. Is the multiverse hypothesis falsifiable? Is there any evidence of the slightest kind that hints at its reality?

For this and other reasons, not every physicist or cosmologist has embraced the multiverse hypothesis. Paul Steinhardt, one of the founders of inflationary cosmology, has said that the multiverse proposal, which he calls the "accidental universe idea," is "scientifically meaningless because it explains nothing and predicts nothing."[5] Steinhardt's skepticism began, he says, when it became clear that inflation required a great deal of fine-tuning to lead to our particular universe.

Defenders of the multiverse hypothesis are apt to say that our existence is prima facie evidence for the existence of other universes that are not so congenial to life. How could ours exist, if these others did not also exist? Max Tegmark, for example, claims that this is a prediction like any other. The multiverse idea was not dreamed up whimsically, he says, but comes from close analysis of inflationary cosmology, along with a variety of other ingredients, including quantum mechanics. The multiverse is a prediction following from more basic assumptions, not merely an assumption in its own right.

That's fair enough, but what next? It is certainly true that the existence of a universe like ours is consistent with the multiverse idea, but that hardly amounts to vindication. Scientific theories, historically speaking, may begin by explaining things that we already know, but they prove their worth by making further predictions and allowing more thorough testing of their consequences. Newtonian gravity, for example, provided a tidy explanation of why the orbits of the planets are ellipses—which, indeed, Kepler had already established. But it went on to say that there should be tiny departures of their orbits from perfect ellipticity because of the small gravitational action of one planet on another. That, as I mentioned in chapter 7, is how astronomers predicted the existence of Neptune. On top of that, Newtonian gravity explains how the moon generates twice-daily tides in the earth's oceans and accounts for the fact that the moon's rotation on its axis, over tens of thousands of years, has come to be synchronized with its orbit so that it always presents the same face to earth.

Granted, the multiverse hypothesis is consistent with the existence of our universe. But does it make any specific predictions beyond that? Does it point to any measurable or observable datum that scientists cannot explain in any other way? Not at all.

The pro-multiverse argument, in other words, is little more than a post hoc rationalization of something we already know. If anything, it reminds me of the ancient hypothesis about gravity: all objects naturally wish to move to the center of the universe; the earth is the center of the universe; therefore all objects fall to the surface of the earth. It's a tidy story, but it doesn't qualify as an explanation.

There is a still more basic issue I want to dig into. We've looked at the search for a final theory uniting quantum mechanics and gravity, and we've now veered into a discussion of the multiverse. These are related but clearly distinct notions. The final theory, whether it emerges from superstring models or something quite different, is the theory that is meant to explain how universes come into existence and how they operate. The multiverse hypothesis, on the other hand, is not a specific theory so much as a general scenario, and is a way of allowing us to think about the multitude of universes that the final theory gives rise to, and the place of our universe among them. Although the idea of the multiverse most often crops up in the course of speculations about the kinds of universes that string theory might engender, it's quite possible that some other basic theory, not based on strings, might turn out to be the final answer, but that the multiverse scenario would still apply.

A difficult question arises: If the final theory of fundamental physics and the multiverse hypothesis are complementary, how much explanatory power should we ask of each one? A few examples will show what I mean.

It's widely believed that three dimensions of space and one of time are just right. In a world of two dimensions, you can't have interesting structures—one line can't cross another without the two merging or one of them breaking. Four space dimensions or more also present structural problems: it's hard or impossible to devise an equivalent of inverse square gravity that would allow stable objects in stable orbits around each other. A universe with two distinct dimensions of time is mathematically imaginable but it would be zany, to say the least—again, it's hard to see how it would allow stable, lasting structures.

In short, if you want a universe that allows long-lasting material objects and rational laws of cause and effect, our present situation is just the ticket. Superstring theories, as we have learned, are based on ten or eleven dimensions, but the extra ones are rolled up tight so we don't notice them. It's entirely possible that in other parts of the multiverse there are universes with, let's say, seven macroscopic space dimensions, but it seems likely that such a universe would be filled with a constant flux of ever-changing physical phenomena, so that long-term stability, and therefore life, would be unimaginable.

The same scenario may play out if something other than superstring theory turns out to be the final theory. If this is the case, then the question of why our world has three dimensions of space and one of time is therefore answered, in a fashion anyway, by the multiverse hypothesis. Other kinds of universe might well exist, but we couldn't live in them.

Now consider another curious fact. In our universe we have three "generations" of basic particles: the electron, electron neutrino, and the up and down quarks are the first; then there's the muon and its neutrino, along with the charm and strange quark; and the tau and tau neutrino plus the top and bottom quarks form the third. These three generations are in essence repetitions of the same arrangement, except with systematically higher masses, so why do we have three and not two or four or fifteen? Does the number three have some deep significance that the final theory will one day explain, or is it merely an accident of our particular place in the multiverse?

Another issue: when the various high-energy symmetries of particle physics break as our universe cools, we end up with four fundamental forces—gravity, electromagnetism, and the strong and weak nuclear forces. Is this pattern significant in some basic way, or could there be plausible, life-supporting universes with more or fewer such forces?

Here's the conundrum: even in the context of the multiverse, we would like to see a final theory that can explain some things about the world we live in, but at the same time no one expects it to explain everything. How, then, are we to know which things demand explanation and which do not? How can we possibly know when just the right amount of finality has been achieved? Taken to an extreme, the multiverse hypothesis says that pretty much every fundamental fact about our universe is up for grabs. If that's so, is it worthwhile for physicists to worry about why there are three generations of particles or four elementary forces? I'm reminded of Bertrand Russell's dig at Thomas Aquinas: if he could find a justification for some important principle in Scripture, all well and good, but if not, well, there's always faith.[6] Researchers in fundamental physics may likewise hope to explain natural phenomena by means of an all-encompassing final theory, but if not, there's always the multiverse.

When physicists hoped to find a single theory that would explain everything, they at least had a grand ambition, if admittedly an implausible one. But now that the multiverse hypothesis serves as a cosmic excuse of last resort for any unexplained fact, the possibility of any sort of empirical vindication for a final theory of fundamental physics is fatally undermined. And that means that researchers in fundamental physics must rely more than ever on principles of mathematical elegance to tell them which theoretical structures are the ones to embrace. As I suggested in chapter 13, however, reliance on aesthetic judgments turns the search for a final theory into a circular proposition. The only useful criterion left for ascertaining whether we have arrived at a final theory is whether it looks right—whether it has mathematical elegance and beauty and whatnot in all the right proportions. If it did not—if it were a complete theory, consistent in all respects, but had too much artifice about it, too much careful adjustment—most physicists would likely think that it wasn't the real deal after all and that further refinement was needed to tidy up loose ends and explain otherwise arbitrary seeming choices. A good theory has to be a good-looking theory—and the people who make that judgment are the scientists trying to create the theory.

String theory, even after all the research hours invested in it, is still a long way from fulfilling its promise. But that doesn't stop its adherents from insisting it represents our best hope. "Good wrong ideas are extremely scarce," Edward Witten has said, "and good wrong ideas that even remotely rival the majesty of string theory have never been seen."[7] Meaning, in other words, that string theory is so wonderful it's highly unlikely to be wrong. Leonard Susskind, asked by the *Economist* how we can know whether string theory is a correct description of nature, responded that string theory and our knowledge of the physical world ought to at least "co-exist with each other. If that's all string theory does then it's still an enormous conceptual advance."[8] Meaning that a mathematical framework that can accommodate, without contradiction, what we understand of the physical world is the most we can expect. Perhaps so—but that seems a long way from the specific, detailed, particular, quantitative empirical agreement between theory and observation that has been the hallmark of traditional science since it was begun by Galileo.

There are options besides string theory, as Lee Smolin describes in his book *Three Roads to Quantum Gravity*. Perhaps one of these alternative avenues will lead to a place in which a theory of quantum gravity will be susceptible to a real experimental test—meaning a critical test, able to say yes or no to the theory, that can be conducted by means of feasible earthbound or astrophysical methods. It seems unlikely, though. The alternatives to string theory that Smolin discusses may have their virtues, but like string theory, they refer to tiny length scales and enormous energies that are beyond our powers of investigation.

Fundamental physics, as it is conducted today, has come down to a matter of juggling mathematical formulations to find structures that agree with what we know of the physical world and then hoping that we can recognize a right answer, or the best answer, when we see it. James Clerk Maxwell, in his Scottish brogue, once observed to his friend P. G. Tait that "heckling of equations through ither" (throwing them around higgledy-piggledy) was no way to solve problems in physics.[9] When the aging Albert Einstein, in a 1933 lecture, suggested that "pure mathematical construction enables us to discover the concepts and the laws connecting them which give us the key to the understanding of the phenomena of Nature,"[10] his biographer Abraham Pais commented, "I cannot believe that this was the same Einstein who had warned Felix Klein in 1917 against overrating the value of formal points of view 'which fail almost always as heuristic aids.' "[11]

Tegmark's mathematical universe hypothesis offers an even more baffling denouement. In his scenario, the multiverse on the largest scale encompasses all mathematical structures permitted by the rules of logic and consistency. Supposing that we ever find a final theory, it will have a mathematical formulation that applies only to some fraction of the multiverse, including (obviously) the part we inhabit. In other parts of the multiverse, there will be final theories based on different mathematical structures. If this is so, even the final theory we find, assuming we can in fact find one, is nothing special, just like everything else about our universe. This appears to mean it will take enormous intellectual effort to uncover the correct final theory for our universe, and when we do, the answer will mean precisely nothing. The mathematical structure of our universe, like everything else, just happens to be.

But will we, in fact, ever be able to settle on a single final theory, or will there be many possibilities, each with pluses and minuses? If the only true test of a final theory is whether it meets intellectual criteria *that we have set,* then the enterprise begins to seem both unending and pointless. We come back, circuitously, to Eugene Wigner's puzzlement over the "unreasonable effectiveness" of mathematics in physical theory. In *The End of Physics,* as I mentioned in chapter 10, I presented the pat answer that "mathematics is the language of science because we reserve the name 'science' for anything that mathematics can handle." I began to think that was too facile an explanation, but now I am not so sure. Mario Livio, in the last chapter of *Is God a Mathematician?,* comes to a similar conclusion, suggesting that the fundamental laws of nature are tidily mathematical because we recognize as fundamental laws only those principles that can be expressed tersely in neat mathematical form. This is a more subtle point than I arrived at thirty years ago: it is not merely, as I thought then, that the so-called hard sciences are defined by their reliance on mathematical equations—rather, it involves the deeper question of what we mean when we talk about fundamental laws of nature. They are precisely the ones that say a lot with a little—that can be printed on a T-shirt, as the late Nobel physicist Leon Lederman liked to say.

Taken to an extreme (and some philosophers of science lean in this direction), this attitude suggests that the fundamental laws of nature are ultimately defined not by nature but by us. But in traditional Galilean science, there's a crucially important proviso: the laws we come up with have to work. They have to account for phenomena we can see and measure, and they have to do so in all relevant circumstances, all the time. This is no easy task. Simple mathematical equations are not hard to come by. Elegant ones are not uncommon either. But simple elegant equations that survive stringent empirical testing to show that they really do capture some essential aspect of the behavior of the natural world are neither common nor easy to identify. Such laws prove themselves, over the centuries, by their universality and their durability. As Galileo first truly understood, experiment and observation form the acid test of a theory and distinguish useless from useful mathematical language.

But when we move beyond experiment and observation, into the realm of fundamental physics, what then? The criteria for a good theory are consistency (no internal contradictions, agreement with what we know of the physical world), and, depending on the eye of the observer, beauty or elegance or brevity or some other combination of characteristics that make a theory appealing. This may well be intellectually strenuous work, but the question looms: Is it science?

In chapter 12, I suggested that the majority of current activities by working scientists are better described as varieties of engineering. They are disciplines that rest on established foundations, in which new fundamental laws that upend the received wisdom are not only unsought but would be highly unwelcome. In these areas of science, the goal is to understand the natural world in ever greater detail and with ever greater precision, according to an accepted set of principles, and also, in many cases, to build devices or invent technologies that capitalize on those principles. Nanoscience, molecular biology, quantum computing, adaptive materials—all engineering. There is also, of course, such continuing work as the discovery of new plant, animal, and insect species, and it makes no sense to call these activities engineering. Such discoveries, though, are entirely consistent with our understanding of biological evolution and do not threaten in any way to contradict the laws of chemistry or quantum mechanics. If novelties arise—a previously unknown chemical signaling system in an insect society, for example—the task is then to understand that phenomenon on the basis of known science.

The one area of modern science that clearly can't be described by the engineering paradigm, and in which the whole point is to uncover new basic laws, is fundamental physics—particle physics, the unification of gravity with quantum mechanics, and cosmology. In this final chapter I am ready to declare that research in this area, no matter its intellectual pedigree and exacting demands, is better thought of not as science but as philosophy. It's philosophy of a very modern style, in that it demands deeply specialized knowledge of mathematics, but it's also philosophy in a very ancient sense, because it presupposes that introspection, driven by logical argument, will suffice to reveal in full the workings of the natural world. Plato and his acolytes thought they could understand the universe by the power of contemplation alone and thus reason out the geometry of the heavens. In the same way, modern researchers into fundamental physics are using logic and reason and the most advanced tools of mathematics in the hope of determining how the physical world works, from the tiniest scales up to the origin of the universe.

The modern researcher, unlike the Platonist of old, is equipped not only with an immensely sophisticated set of intellectual tools but also with a huge amount of quantitative information about subatomic particles and cosmological structure. In *Lost in Math,* Sabine Hossenfelder recounts a conversation with the theorist Nima Arkani-Hamed that touches on exactly this point. Arkani-Hamed suggests that fundamental physics has no real need of novel experiments because "our field is so mature that we get incredible constraints already by old experiments. The constraints are so that they rule out pretty much everything you can try."[12] That's true, to an extent—the search for theoretical structures that can accommodate everything we already know is exceedingly difficult, and the search for theoretical refinements that can potentially extend our understanding without doing damage to everything we already know is harder still. Even so, I don't think the constraints on theory imposed by our current store of empirical knowledge will prove to be enough. Necessarily, the imagination of the theorists must take us into uncharted territory in order to find connections and simplifications that go beyond what has already been proposed. And those extensions will demand new tests. The idea that we can winnow theoretical possibilities down to just a single one is, as I have explained, no longer seen as realistic by the researchers engaged on that quest. But if there are different possibilities, how are we to distinguish between them without new empirical information?

In any case, as I have explained, hopes of pinning down theories in ever more precise ways are at odds with the multiverse idea, which openly acknowledges that there are many possible fundamental theories, that ours is to some extent arbitrary, and that some peculiarities of our universe are just that: peculiarities that apply to our cosmic habitat and no other.

The unanswerable difficulty, as I hope has become clear by now, is that researchers in fundamental physics are exploring a world, or worlds, hopelessly removed from our experience. These worlds have extra dimensions of space, tightly rolled up; strings of energy moving around in those dimensions; symmetries that exist only in those dimensions, or at enormously high energies. The goal is somehow to explain what we know of our tangible world by reference to worlds we cannot see and can never hope to see. What defines those unknowable worlds is perfect order, mathematical rigor, even aesthetic elegance. This is the modern version of Plato's distinction between heaven and earth. In the heavens, all is mathematical perfection; on earth, it's a hopeless, chaotic mess.

Plato never supposed that by explaining the heavens he would also explain the earth, but that's exactly what modern fundamental physics proposes. Find the one true explanation for the high-energy multidimensional perfectly symmetric world of our imagination, and somehow an explanation for our mundane universe will emerge. Exploration of that modern Platonic realm is what research in fundamental physics amounts to, and it is a matter of reason and logic alone, a search for a final theory that provides intellectual rationalization of what we know about the physical world, just as the old Platonic system sought to explain the regularities of the heavens in terms of pure geometry.

Modern science began, remember, when Galileo broke the hold of ancient philosophy, which had in his day been subsumed into the orthodoxy of the Catholic Church. Fundamental physics today is in the grip of its own orthodoxy. In *The Trouble with Physics*, Lee Smolin recounts how pioneers of string theory in the early 1980s were regarded as mavericks and eccentrics. Then, when string theory took off, all the critics were quick to jump on the bandwagon. After that, those who pursued string theory "looked down on those who wouldn't, or (the suggestion was always there) couldn't. Very quickly there developed an almost cultlike atmosphere. You were either a string theorist or you were not."[13] This modern orthodoxy is imposed not from outside science but from within. Both Smolin and Peter Woit, in Woit's 2006 book *Not Even Wrong,* lament that the stranglehold of string theory makes it hard for young researchers to pursue alternatives. Excommunication from the academic world means failing to be awarded research grants or being denied tenure. Still, no one faces lifelong house arrest or exile (although Smolin moved to

Canada to work at the Perimeter Institute for Theoretical Physics, a research foundation outside the formal academic world that was established by the entrepreneur behind the BlackBerry company).

But even the critics of string theory, it seems, continue to believe that fundamental physics, pursued by traditional research methods, will one day yield a satisfactory answer. They would like to see a more diverse range of strategies in the quest for a final theory, but they appear not to doubt that such a thing ought to exist. Sabine Hossenfelder, a disillusioned particle physicist, finishes *Lost in Math* with a plea to her fellow scientists to put aside modern foolishness and return to the time-honored scientific practices of old.

I am skeptical. We are faced with a real question that we cannot resist asking: How did our universe, in the form we perceive, come to be? We can't *not* ask the question. But science cannot answer it. Is our universe a single habitat, one of a kind, that happens to be uniquely suited to our existence? That is scientifically implausible, the modern consensus goes, and has unwelcome theological implications. But then is our universe merely one of many, special to the extent that we are in it, but otherwise run-of-the-mill? That makes most of the questions about why our universe is as it is either unanswerable (it happens to be thus, but need not be) or self-evident (it has to be thus, or we wouldn't be here making inquiries). Either way, we do not reach a satisfying conclusion.

This is why I classify fundamental physics today as a kind of philosophy. The questions it asks are basic and important; they cannot easily be set aside. But the answers it posits are less about a strictly rational understanding of the universe and more about finding a scenario that we deem intellectually respectable. Inevitably, any enlightenment that emerges depends as much on our prejudices about what the right answer ought to look like as on unquestioned matters of empirical fact.

What researchers in fundamental physics are striving for—in vain, I believe—is a rational but idealized system that accommodates the facts but also satisfies whatever aesthetic preferences we bring to bear. William James, I was pleased to discover, had something to say on the search for ideal philosophies: "What the system pretends to be is a picture of the great universe of God. What it is—and oh so flagrantly!—is the revelation of how intensely odd the personal flavor of some fellow creature is."[14]

James was castigating idealism that came in different flavors according to their various philosophical champions. But fundamental physics today suffers from a narrow idealism that arises from within. At the start of this chapter I said that scientists have traditionally not fretted too much about where they are going. Instead, they tackle the problems that stand immediately in front of them and let the future unfold as it will. Science, after all, is the exploration of the unknown. But researchers in fundamental physics, knowingly or not, have adopted entirely the opposite strategy: they have declared in advance what they are looking for and are toiling to create a theory that matches their expectations. They do this, arguably, out of necessity. Observation, experiment, and fact-finding are no longer able to guide them, so they must set their path by other means, and they have decided that pure rationality and mathematical reasoning, along with a refined aesthetic sense, will do the job.

As an intellectual exercise, fundamental physics retains a powerful fascination, at least for those few who are fully able to appreciate it. But it is not science. It's not that I think such research should cease altogether. But I wish its practitioners would take the trouble to ponder where they are going, and to what end.

Acknowledgments

This book is the result of my intermittent cogitation over the nearly three decades since *The End of Physics* appeared. During that time I have badgered many people with my thoughts and I am grateful for their forbearance.

Susan Rabiner, my agent, performed her usual thoughtful and challenging service in pushing me to clarify my purpose as I was getting this project off the ground.

Edward Kastenmeier, my editor at Doubleday, helped me immensely by streamlining an earlier draft of the book and detecting confusions and infelicities great and small as we got closer to a final version.

I am thankful more than I can say for moral support and encouragement from Peggy Dillon.

Notes

PART I HOW SCIENCE BEGAN

My accounts of Galileo and his contemporaries are taken mostly from the two biographies, both titled *Galileo,* by Heilbron and Drake, and I have pointed out some of their differences of interpretation. Gingerich's *The Book Nobody Read* has more on Galileo, as well as on Copernicus and Kepler and the rest. For the specifics of Greek thought as it relates to science, I relied on *Early Greek Science* by G. E. R. Lloyd, while for general commentary on the philosophy of ancient times and the Renaissance era I made use of Russell's entertainingly opinionated *History of Western Philosophy.*

1 Galileo Invents Science

1. Daniel Santbech, *Problematum astronomicorum et geometricorum sectiones septem* (c. 1561), https://commons.wikimedia.org/w/index.php?curid=60700867.
2. Stillman Drake, *Galileo,* 22.
3. G. E. R. Lloyd, *Early Greek Science,* Kindle, chap. 8.
4. Drake, *Galileo,* 25–26.

2 Copernicus Doesn't Quite Invent Astronomy

1. Owen Gingerich, *The Book Nobody Read,* 20.
2. Illustration from "The Retrograde Motion of Planets," ScienceU, http://www.scienceu.com/observatory/articles/retro/retro.html.

3. Gingerich, *The Book Nobody Read*, 32.
4. Steven Weinberg, *To Explain the World*, 162–63.
5. I thank David Goldfrank of Georgetown University for this suggestion.

3 That Old-Time Philosophy

1. Armand Leroi, *The Lagoon*, 183, 138, 227, respectively.
2. G. E. R. Lloyd, *Early Greek Science*, Kindle, chap. 8.
3. Ibid., chap. 3.
4. Ibid.
5. Ibid., chap. 7.
6. Ibid.

4 The Holy Roman Empire Strikes Back

1. Bertrand Russell, *History of Western Philosophy*, 428.
2. Owen Gingerich, *The Book Nobody Read*, 146.
3. Stillman Drake, *Galileo*, 3.
4. Ibid., 64.
5. John L. Heilbron, *Galileo*, 222.
6. Pierre Duhem, *Essays in the History and Philosophy of Science*, 147.
7. See John Polkinghorne, *Science and Religion in Quest of Truth* (New Haven: Yale University Press, 2011), especially chap. 2.
8. Heilbron, *Galileo*, 317.
9. Ibid., 247.

5 How Science Uses Mathematics

1. Mark A. Peterson, *Galileo's Muse*, 167.
2. Ibid., 18.
3. Galileo Galilei, *Dialogues Concerning Two New Sciences*, Kindle, Second Day.
4. John L. Heilbron, *Galileo*, 142.
5. M. A. Hoskin, "Nature and Mathematics," in *The Making of Modern Science: Six Essays by Gerd Buchdahl, M. A. Hoskin, A. Rupert Hall, Marie Boas Hall, Sam Lilley, Charles Raven*, ed. A. Rupert Hall (Leicester, UK: Leicester University Press, 1960), 17.
6. Stillman Drake, *Galileo*, 17.

PART II CLASSICAL SCIENCE REIGNS SUPREME

Westfall (*Never at Rest*) and Gleick (*Isaac Newton*) are my sources for the life and work of Newton. My discussion of the rise of thermodynamics and electromagnetic theory rely greatly on my books *Degrees Kelvin* and *Boltzmann's Atom,* respectively.

6 Mastery of Motion

1. Galileo Galilei, *Dialogues Concerning Two New Sciences,* Kindle, Third Day.
2. Richard S. Westfall, *Never at Rest,* 403 et seq.
3. James Gleick, *Isaac Newton,* 3.

8 The Limits of Pragmatism

1. For the full story see Clifford Truesdell's *Tragicomical History of Thermodynamics, 1822–1854,* Studies in the History of Mathematics and Physical Sciences 4 (New York: Springer-Verlag, 1980). This is not a book for the faint of heart.
2. For the full story see David Lindley, *Boltzmann's Atom.*
3. Pierre-Simon de Laplace, *A Philosophical Essay on Probabilities* (New York: Dover, 1951), 4.
4. See https://upload.wikimedia.org/wikipedia/commons/5/57/Magneto873.png.
5. Emilio Segrè, *From X-Rays to Quarks: Modern Physicists and Their Discoveries* (San Francisco: W. H. Freeman, 1980), 2.
6. David Lindley, *Degrees Kelvin,* 202.

PART III FUNDAMENTAL PHYSICS CHARTS ITS OWN COURSE

The history of particle physics in the twentieth century, preceding the outbreak of string fever, is ably recounted by Crease and Mann in *The Second Creation.* Pais's *Inward Bound* is a thorough but much more technical account. I also told part of the story in *The End of Physics.* For more recent developments, especially string theory and its kin, I worked mostly from Greene's books *The Elegant Universe* and *The Hidden Reality.*

9 Dirac Invents Antimatter

1. P. A. M. Dirac, "Quantised Singularities in the Electromagnetic Field," *Proceedings of the Royal Society A* 133 (September 1, 1931): 60, https://doi.org/10.1098/rspa.1931.0130.

10 Wigner's Enigmatic Question

1. See Robert P. Crease and Charles C. Mann, *The Second Creation,* chaps. 14 and 15 especially.
2. E. P. Wigner, "The Unreasonable Effectiveness of Mathematics in the Natural Sciences," *Communications on Pure and Applied Mathematics* 13 (February 1960): 1–14, https://doi.org/10.1002/cpa.3160130102.
3. Crease and Mann, *The Second Creation,* 266.
4. David Lindley, *The End of Physics,* 4.
5. Bertrand Russell, *An Outline of Philosophy* (1927; repr., London: Routledge Classics, 2009), chap. 15.
6. Alice Calaprice, ed., *The Quotable Einstein,* 197.
7. Max Planck, *Where Is Science Going?* (1932; repr., Woodbridge, CT: Ox Bow Press, 1981), 214.

11 All This Useless Beauty?

1. Paul Dirac, "The Evolution of the Physicist's Picture of Nature," *Scientific American,* May 1963, https://blogs.scientificamerican.com/guest-blog/the-evolution-of-the-physicists-picture-of-nature/.
2. Ibid., 396.
3. See David Lindley, *Degrees Kelvin,* 264 et seq.
4. Quotations from Oliver Lodge and Albert Michelson, respectively, quoted in Sabine Hossenfelder, *Lost in Math: How Beauty Leads Physics Astray.*
5. J. J. Thomson, *Recollections and Reflections* (London: Bell & Sons, 1936), 95.
6. David Lindley, *Boltzmann's Atom,* 52.
7. Bertrand Russell, "The Study of Mathematics," *New Quarterly* (November 1907), http://bertrandrussellsocietylibrary.org/br-pe/br-pe-ch3.html.
8. P. A. M. Dirac, "The Quantum Theory of the Electron," *Proceedings of the Royal Society A* 117 (February 1, 1928): 610–24, https://doi.org/10.1098/rspa.1928.0023.
9. P. A. M. Dirac, "The Relation Between Mathematics and Physics," *Proceedings of the Royal Society of Edinburgh* 59 (1938–39), Part II: 122–29, http://www.damtp.cam.ac.uk/events/strings02/dirac/speach.html.

10. G. H. Hardy, *A Mathematician's Apology*, 119.

11. Ibid., 131.

12. Galina Weinstein, "George Gamow and Albert Einstein: Did Einstein Say the Cosmological Constant Was the 'Biggest Blunder' He Ever Made in His Life?" arXiv (October 3, 2013), https://arxiv.org/ftp/arxiv/papers/1310.1033.pdf.

13. Farmelo, *The Strangest Man*, 428.

14. David Lindley, *Uncertainty*, 188.

15. John Schwarz, "Resuscitating Superstring Theory," *The Scientist* (November 16, 1987), https://www.the-scientist.com/research/resuscitating-superstring-theory-63323.

PART IV SCIENCE OR PHILOSOPHY?

This final section draws on many of the ideas presented earlier along with numerous other sources cited as appropriate.

13 The Last Problems

1. Ian Stewart, *Why Beauty Is Truth*, 247.

2. Brian Greene, *The Hidden Reality*, 96.

3. Stephen Hawking, *A Brief History of Time: From the Big Bang to Black Holes* (New York: Bantam: 1988), 175.

4. "Using Maths to Explain the Universe," Prospero, *Economist*, July 2, 2013, https://www.economist.com/prospero/2013/07/02/using-maths-to-explain-the-universe.

14 The Byte-Sized Universe

1. *Proceedings of the Royal Society of Edinburgh* (April 19, 1852); see also David Lindley, *Degrees Kelvin*, 108–9.

2. See David Lindley, *Boltzmann's Atom*, chaps. 6 and 7.

3. Seth Lloyd, *Programming the Universe*, 187.

15 Is Math All There Is?

1. Mario Livio, *Is God a Mathematician?*, 229.

2. Max Tegmark, *Our Mathematical Universe*, Kindle, chap. 7.

3. Ibid., chap. 10.

4. Ibid., chap. 7.

5. Ibid., chap. 11.

16 The Dream Universe

1. Michio Kaku, *Parallel Worlds: A Journey Through Creation, Higher Dimensions, and the Future of the Cosmos* (New York: Doubleday, 2005), 282.
2. Max Tegmark, *Our Mathematical Universe*, Kindle, chap. 6.
3. Martin Rees, *Just Six Numbers*, 10.
4. See Brian Greene, *The Hidden Reality*, chap. 6; also Sabine Hossenfelder, *Lost in Math*, chap. 5.
5. John Horgan, "Physicist Slams Cosmic Theory He Helped Conceive," *Scientific American, Cross-Check*, Dec. 1, 2014, https://blogs.scientificamerican.com /cross-check/physicist-slams-cosmic-theory-he-helped-conceive/.
6. Bertrand Russell, *History of Western Philosophy*, 453.
7. Rees, *Just Six Numbers*, 161.
8. "Using Maths to Explain the Universe," Prospero, *Economist*, July 2, 2013, https:// www.economist.com/prospero/2013/07/02/using-maths-to-explain-the-universe.
9. Cargill Gilston Knott, *The Life and Scientific Work of Peter Guthrie Tait* (Cambridge, UK: Cambridge University Press, 1911), 117.
10. A. Einstein, "On the Method of Theoretical Physics," *Herbert Spencer Lecture* (Oxford, UK: Oxford University Press, 1933).
11. Abraham Pais, *Subtle Is the Lord: The Science and Life of Albert Einstein* (Oxford, UK: Oxford University Press, 1982), 347.
12. Sabine Hossenfelder, *Lost in Math*, chap. 4.
13. Lee Smolin, *The Trouble with Physics*, 116.
14. William James, "Lecture 1," in *Pragmatism: A New Name for Some Old Ways of Thinking* (1907; repr., New York: Dover, 2018).

Selected Bibliography

Calaprice, Alice, ed. *The Quotable Einstein*. Princeton, NJ: Princeton University Press, 1996.

Crease, Robert P., and Charles C. Mann. *The Second Creation: Makers of the Revolution in Twentieth-Century Physics*. New York: Macmillan, 1986.

Drake, Stillman. *Galileo*. Oxford, UK: Oxford University Press, 1980.

Duhem, Pierre. *Essays in the History and Philosophy of Science*. 1892–1915. Translated and edited by Roger Ariew and Peter Barker. Indianapolis: Hackett, 1996.

Farmelo, Graham. *The Strangest Man: The Hidden Life of Paul Dirac, Mystic of the Atom*. New York: Basic Books, 2009.

Galilei, Galileo. *Dialogues Concerning Two New Sciences*. 1638. Translated by Henry Crew. Overland Park, KS: Digireads.com, 2010.

Gingerich, Owen. *The Book Nobody Read: Chasing the Revolutions of Nicolaus Copernicus*. 2004. Reprint, New York: Walker, 2005.

Gleick, James. *Isaac Newton*. 2003. Reprint, New York: Vintage Books, 2004.

Greene, Brian. *The Elegant Universe: Superstrings, Hidden Dimensions, and the Quest for the Ultimate Theory*. New York: W. W. Norton, 1999.

———. *The Hidden Reality: Parallel Universes and the Deep Laws of the Cosmos*. New York: Alfred A. Knopf, 2011.

Hardy. G. H. *A Mathematician's Apology*. Cambridge, UK: Cambridge University Press, 1940. Reprint, Cambridge, UK: Canto, 1992.

Heilbron, John L. *Galileo*. Oxford, UK: Oxford University Press, 2010.

Heller-Roazen, Daniel. *The Fifth Hammer: Pythagoras and the Disharmony of the World*. New York: Zone Books, 2011.

Hossenfelder, Sabine. *Lost in Math: How Beauty Leads Physics Astray*. New York: Basic Books, 2018.

Leroi, Armand. *The Lagoon: How Aristotle Invented Science.* London: Bloomsbury, 2014.

Lindley, David. *Boltzmann's Atom: The Great Debate That Launched a Revolution in Physics.* New York: Free Press, 2001.

———. *Degrees Kelvin: A Tale of Genius, Invention, and Tragedy.* Washington, D.C.: Joseph Henry Press, 2004.

———. *The End of Physics: The Myth of a Unified Theory.* New York: Basic Books, 1993.

———. *Uncertainty: Einstein, Heisenberg, Bohr, and the Struggle for the Soul of Science.* New York: Doubleday, 2007.

Livio, Mario. *Is God a Mathematician?* New York: Simon & Schuster, 2009.

Lloyd, G. E. R. *Early Greek Science: Thales to Aristotle.* London: Chatto & Windus, 1970. Kindle.

Lloyd, Seth. *Programming the Universe: A Quantum Computer Scientist Takes on the Cosmos.* New York: Alfred A. Knopf, 2006.

Pais, Abraham. *Inward Bound: Of Matter and Forces in the Physical World.* Oxford, UK: Oxford University Press, 1986.

Peterson, Mark A. *Galileo's Muse: Renaissance Mathematics and the Arts.* Cambridge, MA: Harvard University Press, 2011.

Rees, Martin J. *Just Six Numbers: The Deep Forces That Shape the Universe.* New York: Basic Books, 2000.

Russell, Bertrand. *History of Western Philosophy.* London: George Allen and Unwin, 1946.

Smolin, Lee. *The Trouble with Physics: The Rise of String Theory, the Fall of a Science, and What Comes Next.* New York: Houghton Mifflin, 2006.

Stewart, Ian. *Why Beauty Is Truth: A History of Symmetry.* New York: Basic Books, 2007.

Tegmark, Max. *Our Mathematical Universe: My Quest for the Ultimate Nature of Reality.* New York: Alfred A. Knopf, 2014.

Weinberg, Steven. *To Explain the World: The Discovery of Modern Science.* New York: Harper, 2015.

Westfall, Richard S. *Never at Rest: A Biography of Isaac Newton.* Cambridge, UK: Cambridge University Press, 1980.

Wootton, David. *The Invention of Science: A New History of the Scientific Revolution.* New York: Harper, 2015.

Index

acceleration, 60–61, 65–66, 68
action-at-a-distance theories,
    89–92, 94
alchemy, 66
algebra, 60, 68, 69, 75, 77
    matrix mechanics and, 103, 104,
    117–18, 128, 160
Ampère, André-Marie, 91
Anderson, Carl, 108
antielectron:
    discovery of, 106–8, 110, 114, 183
    Wigner's theorem and, 112–13
antimatter, 109
applied mathematics, 78, 128
Aquinas, Thomas, 38, 40, 192–93
    Summa contra gentiles, 37
Arabic world, 34
    algebra and, 60
Archimedes, 7, 64, 69
Aristarchus, 17
Aristotelian natural philosophy, 4–6, 9,
    15, 16, 19, 21, 23, 24, 32, 41, 62
    Catholic orthodoxy and, 37, 38, 48
    idea of purpose and, 11
    motion and, 3–6, 9, 10, 11, 24, 37, 38
    in pre-Renaissance Europe, 25, 35
Aristotle, 4, 6, 10, 19, 25–27, 35, 37, 56,
    117, 143, 183
    animal life analyzed by, 26–27

On the Heavens, 10, 32
    on physics and astronomy, 27, 32
Aristoxenus, 52
Arkani-Hamed, Nima, 197–98
artificial intelligence, 142
astrology, 8, 14–15
astronomy, 1, 13–24, 170, 186
    Aristotle's theories on, 27, 32
    Plato's contributions to, 29–31
    Ptolemaic system and, 9, 13–18, 21–22,
    32, 44
    see also heliocentrism
atoms, 96, 99, 102, 135, 136, 157–58, 180
    kinetic theory of heat and, 86–88, 96,
    97, 104, 115, 123
    Newtonian billiard ball trope and, 175
    nuclei of, 149
    vortex model of, 123, 127
Augustine, Saint, 40
Averroës, 34–35, 37, 38
    The Incoherence of the Incoherence, 35

Babylonians, 14–15, 28
Barrow, Isaac, 62, 66
beam, analysis of bending of, 72–73
Beckmann, Petr, *A History of Pi,* 172
Becquerel, Henri, 97
Bellarmine, Cardinal Robert, 42, 43,
    44, 46

Bell Labs, 166
Bernoulli, Daniel, 78
Bessel, Friedrich Wilhelm, 78
Bessel functions, 78–79, 80, 81, 120
beta radioactivity, 149
Bible, 38–39, 40, 42, 43, 47, 48, 66
big bang cosmology, 132, 152, 153, 154,
    164, 165
biochemistry, 136
bioengineering, 134
Biot, Jean-Baptiste, 91
Biot-Savart law, 91
Bohr, Niels, 131, 160
Boltzmann, Ludwig, 86–87, 123–24,
    163–64, 166
bosons, 154–55
botany, 141–42
Brahe, Tycho, 23

calculus:
    differential equations in, 65, 74,
        75–80, 93, 127–28
    integral, 65, 73, 127–28
    Newton's invention of, 63–65, 67,
        68–70, 71, 115–16
    Newton's *Principia Mathematica* and,
        68–69, 90
    versatility of, 71–73
calendrical calculations, 13, 14, 23
cannonball, flight of, 3–6, 11, 38, 47, 50,
    53, 114
Carnot, Sadi, 85–86, 162
Cartesian coordinates, 62, 63
Catholic Church, 1, 6, 8, 9, 18, 24, 25,
    35–48, 199
    Galileo's strife with, over
        heliocentrism, 24, 25, 38–48
    intellectual foundation for orthodoxy
        of, 37–39
    power of, 36–37, 39–40, 42
CERN, 150–51, 155
Chadwick, James, 110
charge (quantum property), 112–13

charm (quantum property), 113
charm quark, 192
Châtelet, Émilie du, 90
chemistry, 96, 136, 140
circles, 12, 49
    calculating area of, 64
    heavenly paths believed to be, 9, 14,
        15–16, 17, 30–31, 32, 70
    as special case of ellipse, 66, 67
climate change, 134, 141
cold fusion, 139–40
computers:
    calculations of, 180–81
    quantum, universe seen as, 168–71
    quantum vs. traditional, 167–68
    solving equations with, 80–81
consciousness, question of, 142
conservation of energy, 162
Copernican principle, 187–88
Copernicus, Nicolaus, 13–14, 16–18, 62
    De revolutionibus, 13, 14, 18, 39, 40
    heliocentric system hypothesized by,
        6, 11, 13–14, 17–18, 19, 21, 23, 38, 39,
        43, 44, 45–47, 187–88
    Little Commentary, 17
cosmological constant, 129–30, 131, 152,
    155, 188
cosmology, xii, xiii, xiv, 20, 37, 41, 135,
    162, 200
    big bang and, 132, 152, 153, 154, 164,
        165
    God question and, 183, 186, 187
    inflationary model of, 152–54
    Platonism and, 50
    quantum computing and, 168–71
Coulomb, Charles-Augustin de, 91
Council of Trent (1545–1563), 37–38, 40
Crease, Robert, and Charles Mann, *The
    Second Creation*, 111

dark energy, 130, 188
Darwin, Charles, 97
Descartes, René, 62, 63, 90, 91, 92

differential calculus, 65, 74, 93, 127–28
tools for solving equations in, 75–80
dimensions of space and time:
extra, suggestions of, 138, 139, 147, 148, 155, 156, 157, 159, 175, 191–92, 198
optimal (three dimensions of space and one of time), 191–92
Dirac, Paul, 101–2, 104–9, 160, 177
antielectron discovery and, 106–8, 110, 114, 183
electron equation devised by, 105–6, 114, 124–27
magnetic monopole predicted by, 131
mathematical beauty favored over empirical accuracy by, 122, 129, 131
DNA, 138, 140
Drake, Stillman, 41, 42–43, 55
drumskin, vibrations of, 78–79

earth, believed to be in center of universe, 9, 15–16, 24, 29, 31, 42–43, 44
Economist, 159
Einstein, Albert, 102, 105, 120, 121, 122, 137, 160, 183–84, 194
cosmological constant and, 129–30, 152
general relativity and, 147, 154, 160, 179, 183–84
God question and, 156, 183, 186
electromagnetic field, 94, 95, 96, 97, 104, 147, 150, 155, 159
physical reality and, 176, 177
electromagnetic force, 186
unification of weak force and, 150–51
electromagnetic waves, 148
electromagnetism, 91–96, 97, 104, 131, 154, 182, 192
action at a distance and, 91
particle associated with, 150
unified theory of gravity and, 147, 148–49

electron neutrino, 192
electrons, 103, 135, 136, 155, 175, 184, 192
Dirac's equation for, 105–6, 114, 124–27
discovery of antielectron and, 106–8, 110, 114, 183
Kaluza-Klein theory and, 148–49
spin of, 175
Wigner's theorem and, 112–13
electrotonic state, 176
electroweak force, 154
unification of strong force and, 151–52
ellipse, 75
circle as special case of, 66, 67
as path of planetary orbits, 23–24, 66, 67, 74–75, 185, 190
energy, 148
conservation of (first law of thermodynamics), 86, 97, 102, 162
engineering, 134–43, 173, 196–97
distinguishing science from, 135, 140–43
entomology, 141–42
entropy, 162–64, 165, 166, 176
informational, 168–69
second law of thermodynamics and, 85, 97, 162, 165
epicycles and equants, in Ptolemaic system, 16, 17
episteme-techne distinction, 142–43
Euclid, 7, 9, 62
Eudoxus, 27, 31–32, 37
Euler, Leonhard, 82
evolution, 97

Fahrenheit, Daniel, 83–84
Faraday, Michael, 91–95, 96, 159, 160, 176
Farmelo, Graham:
It Must Be Beautiful: Great Equations of Modern Science, 173
The Strangest Man, 101

Ferdinand, Archduke of Holy Roman
Empire, 23
Fermi, Enrico, 110
fermions, 154–55
Feynman, Richard, 126
"First Three Minutes" of cosmological
history, 154
Fleischmann, Martin, 139
force, in Newton's laws of motion,
65–66, 71
formalism, 174
Fourier, Jean-Baptiste-Joseph, *Théorie
analytique de la chaleur,* 84
Fourier series, 84
fundamental physics, xi, 135, 137, 139,
143, 145, 186
current goals of, xii, 201
evolved into version of philosophy,
xiv, 197–201
grandiose ambitions in, 183–84
increasingly removed from direct
experience, xiv, 198–99
see also cosmology; particle physics;
quantum theory

Galileo, Vincenzo, 7, 8, 50, 51–52, 53
Galileo Galilei, xiii–xiv, 1, 6–12, 13,
18–19, 24, 57–58, 62, 66, 70, 83, 132,
135, 137, 138, 143, 161, 173, 175, 182,
194, 196, 199
The Assayer, 44, 45, 47–48, 54
astronomical observations of, 20–23,
41–42
Catholic Church's strife with, 24, 25,
38–48
*Dialogue Concerning the Two Chief
World Systems, Ptolemaic and
Copernican,* 44, 45–46, 53
*Dialogues Concerning Two New
Sciences,* 6, 10, 47, 53–54, 59–61
hampered by lack of mathematical
tools, 59–61
on mathematics as language of
universe, 47–48, 54, 55–56, 81, 120,
175
mathematics' value and utility as
viewed by, xiii–xiv, 20, 22, 48,
50–56, 122
motion and mechanics studied by,
6, 7, 9–11, 12, 22, 38, 47, 50, 53, 55,
59–61, 63, 65, 68, 71–72, 114, 116
Platonism erroneously ascribed to,
48, 54–56
Sidereus nuncius, 22
The Starry Messenger, 22, 41
Gell-Mann, Murray, 114–15, 118–19
general relativity, 129, 132, 137, 138, 147,
154, 160, 179
genetic engineering, 134
geo-engineering, 134
geology, 96–97, 136
geometry, xiii, 3–4, 7, 9, 48, 71
Copernicus's work in, 13
curvature of space-time and, 137, 147,
160
Galileo's studies of motion and
mechanics and, 55, 60, 61
Newton's innovations expressed in
form of, 68–69
see also Platonic solids
al-Ghazali, *The Incoherence of the
Philosophers,* 35
Gingerich, Owen, 14, 39–40
The Book Nobody Read, 18
Gleick, James, 68
gluons, 136, 150, 151
God, 20, 35, 49, 183, 185, 186, 187, 201
Copernicanism and, 45–46
planetary orbits and, 14
power of Catholic Church and, 36
golden ratio, 174
grand unified theories, 151–54
Guth's inflationary universe and,
152–54

graviton, 155
gravity, 149, 151, 154, 160, 185, 186, 192
  action at a distance and, 89–91
  ancient hypotheses about, 190
  entropy and, 165
  Newton's inverse square law of, 66,
    67, 69–70, 89–90, 114, 116–17, 120,
    138
  of one planet on another, 72
  quantum, 194
  quantum computers and, 168–69
  of sun, planetary orbits and, 74, 89
  unification of quantum mechanics
    and, 132, 184, 190–91
  unified theory of electromagnetism
    and, 147, 148–49
Greeks, ancient, xiii, 1, 25–33, 34, 174
  mathematics as viewed by, 28–31,
    32–33
  motion as interpreted by, 3–5, 59
  see also Aristotelian natural
    philosophy; specific philosophers
Greene, Brian, 156–57
  The Elegant Universe, 156, 157
  The Hidden Reality, 157
Gregory IX, Pope, 37
ground state, 152–53
groups and symmetries, mathematics
  of, 112–16, 118–19, 128
Guth, Alan, 152–54, 187

Halley, Edmond, 67–68, 74
Hardy, G. H., A Mathematician's
  Apology, 127–29
harmonies:
  cosmic scheme of harmoniousness
    and, 28, 29, 30–31
  musical, 28, 30, 50–53
heat, 83–88
  flow of, through materials, 84, 85
  kinetic theory of, 86–88, 96, 97, 104,
    115, 123

  measurement of, 83–84
  steam engine and, 84–85
  see also thermodynamics
heat death of the universe, 163
Heilbron, John, 43, 44, 45, 46, 54
Heisenberg, Werner, 103, 113, 115, 117–18,
  119, 159–60
heliocentrism, 6–7, 17–20
  Copernicus's model of, 6, 11, 13–14,
    17–18, 19, 21, 23, 38, 39, 43, 44,
    45–47, 187–88
  Galileo's strife with Catholic Church
    and, 24, 25, 38–48
  Kepler's proposal of elliptical
    planetary orbits and, 23–24
  Kepler's views on, 18–19, 24, 39
  theologians' 1616 report on, 42–43, 46
  verses from Joshua and, 38–39, 40
Heller-Roazen, Daniel, The Fifth
  Hammer, 52
Henry, Joseph, 91
heresy, 6, 35, 37, 38
  heliocentrism and, 42, 43, 44, 45,
    46–47
Higgs boson, 149–51
  revealed at CERN's Large Hadron
    Collider, 149–51, 155
Higgs field, 151
Higgs mechanism, 149–52
  electroweak unification and, 151–52
  inflationary universe and, 152–54
Holy Roman Empire, 36–48
Hossenfelder, Sabine, Lost in Math,
  122–23, 155, 197–98, 199
Hubble, Edwin, 129, 130

Index Librorum Prohibitorum, 38, 39, 40
infinite series, 77, 78
inflationary universe, 152–54, 189–90
information:
  growth of, 166–71
  quantitative treatment of, 166

Inquisition, 6, 37, 42, 44, 45, 46–47
  Galileo's espousal of Copernicanism
    and, 42, 44, 45, 46–47
integral calculus, 65, 73, 127–28
intuitionism, 174
inverse square law of gravity, 66, 67,
  69–70, 89–90, 114, 116–17, 120, 138
Islam, 35
isospin, 113–14

James, William, 201
Joshua, 38–39, 40
Jupiter, 14, 20
  moons of, 21, 22, 23

Kaku, Michio, *Parallel Worlds*, 184–85
Kaluza, Theodor, 147, 148
Kaluza-Klein theory, 148–49, 154, 155
Kelvin, William Thomson, Lord, 95, 123,
  135, 162–63
Kepler, Johannes, 18–21, 62
  Astronomia nova, 23, 39
  Brahe's observations analyzed by, 23
  De stella nova, 20–21
  elliptical planetary orbits and, 23–24,
    75, 185, 190
  heliocentrism and, 18–19, 24, 39
  Mysterium cosmographicum, 18–19
  new star observed by, 20–21
  solar system of nested Platonic solids
    pictured by, 19–20, 28, 122, 185
al-Khwarizmi, 34
Klein, Felix, 194
Klein, Oskar, 148
Koyré, Alexandre, 54–55
Kuhn, Thomas, *The Structure of
  Scientific Revolutions*, 137–38

Laplace, Pierre-Simon, marquis de,
  88–89
Large Hadron Collider (LHC), 150–51,
  152, 155
Lederman, Leon, 115, 196

Leibniz, Gottfried, 67, 69
Leroi, Armand, *The Lagoon*, 25–26, 27
life, conditions in universe for, 186–87,
  188–89, 191–92
light, 94, 102, 123, 148
  Newton's ideas on behavior and
    properties of, 66, 68, 82
  speed of, 137
  wave nature of, 82
Lindley, David:
  The End of Physics, xi, xii, 119, 195
  The Myth of a Unified Theory, xi–xii
Livio, Mario:
  The Golden Ratio, 172
  Is God a Mathematician?, 173–74,
    195
Lloyd, G. E. R., 28
Lloyd, Seth, *Programming the Universe*,
  166–69
Luther, Martin, 37

Mach, Ernst, 87–88
magnetic monopoles, 131
Mann, Charles, and Robert Crease, *The
  Second Creation*, 111
Mars, 14, 16, 20
materials science, 136, 141
mathematics, xiii–xiv, 1, 49–56, 71–81,
  159–60, 172–81, 185, 194
  applied, 78, 128
  Arabic world's contributions to, 34
  beauty's significance in, 120–21,
    122–33, 151, 155, 161, 172–73, 183,
    193, 196
  Bessel functions and, 78–79, 80, 81,
    120
  best and purest, marked by its
    uselessness, 127–28
  computer-generated solutions and,
    80–81
  defining, 173–74
  fundamental laws of nature and,
    195–96

Galileo hampered by his lack of,
 59–60
Galileo's attitude toward value and
 utility of, xiii–xiv, 20, 22, 48, 50–56,
 122
gulf between theory and observation
 and, xiv, 107–8, 173–77
Hardy's *A Mathematician's Apology*
 and, 127–29
Hossenfelder's *Lost in Math* and,
 122–23, 155, 197–98, 199
idealized, in Plato's philosophy,
 29–30, 48, 49–50, 52
increasingly abstract nature of, 102,
 104, 113–14, 119
as language in which physical ideas
 are conveyed, but not the *origin* of
 those ideas, 53–54, 81, 94
as language of universe, Galileo's
 dictum on, 47–48, 54, 55–56, 81,
 120, 175
matrices and, 103, 104, 117–18, 128,
 160
philosophers of pre-Christian era
 and, 28–31, 32–33
pure, Wigner's question about
 physicists' use of, 116–21, 173, 195
Pythagoreans' enthusiasm for,
 28, 29
science's relationship to, in age of
 Newton, 74–75
sine curves and, 75–79, 80, 82
string theory and, 159–61, 175–76
of symmetries and groups, 112–16,
 118–19, 128
Tegmark's mathematical universe
 hypothesis and, 173–81, 195
and tools for solving equations in
 classical era, 75–80, 81
see also algebra; calculus; geometry;
 trigonometry
matrix mathematics, 103, 104, 117–18,
 128, 160

Maxwell, James Clerk, 86–87, 94–96, 97,
 104, 131, 148, 176, 182, 194
mechanical view of natural world,
 88–89, 95, 96–97, 140, 142, 175
medical science, 140
mediocrity, principle of, 188
Mercury, 14, 20, 24
mesons, 110–11, 115
method of exhaustion, 64
molecular biology, 138, 140
moon (of earth):
 biblical mentions of, 38, 39, 42
 daily tides in earth's oceans and, 66,
  190
 harmony of the spheres and, 28
 irregularities at edge of, 42
 motion of, 4, 6, 8, 9, 11, 13, 14, 15–16,
  21, 31, 148
moons, 29
 of Jupiter, 21, 22, 23
motion, 59–70
 acceleration and, 60–61, 65–66, 68
 Aristotelian dogma on, 3–6, 9, 10, 11,
  24, 37, 38
 Cartesian coordinates and, 62, 63
 continuous but variable, Newton's
  analysis of, 63–65
 Galileo's studies of, 6, 7, 9–11, 12, 22,
  38, 47, 50, 53, 55, 59–61, 63, 65, 68,
  71–72, 114, 116
 as interpreted by ancient Greeks,
  3–5, 59
 Newton's invention of calculus and,
  63–65, 67, 68–70, 71
 Newton's laws of, 65–66, 68, 69, 71
 see also planetary orbits
multidimensional universe, 138, 161
multiverse hypothesis, xii, 158, 187–93,
 195, 198
muon (originally called mu-meson),
 111, 192
muon neutrino, 192
musical harmonies, 28, 30, 50–53

nanoengineering, 134
Neoplatonism, 37
Neptune, 72, 190
neutrino, 110, 149
neutrons, 107, 110, 113, 135–36, 184
    quark theory and, 114–15
Newton, Isaac, 61–70, 72, 101, 132, 137,
    138, 147, 148, 164–65, 173, 185, 187,
    190
    behavior and properties of light
        studied by, 66, 68, 82
    calculus invented by, 63–65, 67,
        68–70, 71, 115–16
    gravitational law devised by, 66, 67,
        69–70, 89–90, 114, 116–17, 120, 138
    laws of motion formulated by, 65–66,
        68, 69–70, 71
    Opticks, 68, 69
    Principia Mathematica, 68–69, 90
Newtonian mechanics, 69–70, 97, 104,
    175
    ascendance and durability of, 88–89,
        138
    billiard ball trope in, 103, 111, 175
    quantum mechanics in relation to,
        103
    versatility of, 71–74
    see also calculus
nuclear physics, 135–36
numbers, mysticism and, 28
number theory, 128

oceanography, 185–86
Olson, Richard G., *Science and
    Religion*, 43
Ørsted, Hans Christian, 91
Osiander, Andreas, 39

Pais, Abraham, 194
parabola, Galileo's recognition of, in
    cannonball's path, 12, 50, 53, 114
parity (quantum property), 113

particle physics, xii–xiii, xiv, 101, 110–18,
    132, 136, 192
    Dirac's prediction of new particle
        and, 101–2, 104–9, 110
    as philosophy rather than science,
        197–201
    quark theory and, 114–15, 116
    symmetry and group theory in,
        112–16, 118–19
    unification program and, 147–61,
        184, 192–93; *see also* multiverse
        hypothesis; string theory
    use of strange new mathematics in,
        115–21
    Wigner's theorem and, 112–14
particles, 135–36, 175, 179, 180
    classifying by their properties, 113–14
    described in terms of quantum wave
        function, 103, 104–5, 111
    fundamental forces associated with,
        150
    new, proliferation of, 110–11, 113, 149
    quantum spin of, 105–6, 112
    three "generations" of, 192
Paul V, Pope, 41, 42
Pauli, Wolfgang, 110, 127
pendulum, as timer, 11
Perimeter Institute for Theoretical
    Physics, 200
Peterson, Mark, *Galileo's Muse*, 51,
    52, 53
philosophy, fundamental physics
    evolved into version of, xiv,
    197–201
photons, 150, 155
physical quantities, 177–79
    scalars or vectors, 177
    tensors, 178–79
pi, histories of, 172
Planck, Max, 102, 120, 164
Planck mass, 149
planetary astronomers, 186

planetary orbits, 4, 8, 13, 14, 15–16, 17, 20,
21, 30–32, 70, 185, 187, 190
believed to be circular, 9, 14, 15–16, 17,
30–31, 31, 32, 70
elliptical paths of, 23–24, 66, 67,
74–75, 185, 190
gravity of other planets and, 72
planets, 29
harmony of the spheres and, 28
see also heliocentrism; specific
planets
Plato, 19, 27, 37, 52, 54, 117, 143, 161, 173,
183, 184, 185, 199
celestial architecture and, 29–31
empirical investigation disdained
by, 26
idealized mathematics and, 29–30,
48, 49–50, 52
musical harmonies and, 30, 52
Republic, 30
Timaeus, 19
Platonic solids, 49
classical elements associated with,
19, 30
nested, in Kepler's picture of solar
system, 19–20, 28, 122, 185
Platonism, 49–50, 53, 173–74, 185
erroneously ascribed to Galileo, 48,
54–56
Plato's cave, 157
plausibility, criteria for, 170–71
Polkinghorne, John, 43
Pons, Stanley, 139
Popper, Karl, 136–38, 139
pragmatism, 73–74, 82–97
electromagnetism and, 91–96, 97, 104
heat and, 83–88, 96, 97, 104
Protestantism, 6, 37
protons, 106, 107, 110–11, 113, 135–36, 184
quark theory and, 114–15
Ptolemaic astronomical system, 9,
13–18, 21–22, 32, 44

Ptolemy, Claudius, Almagest, 15–16
Pythagoras, 27–28, 69
musical harmonies and, 50–52
Pythagoreans, 27–29, 30, 51
Pythagorean theorem, 28

quanta, 102, 164
quantum computing, 141, 167–68, 197
universe seen as, 168–71
quantum gravity, 194
quantum information, 166
quantum mechanics, 99, 103–5, 128, 132,
137, 138, 141, 159
commonsense visualization of little
help in, 104
Dirac's equation for electron and,
105–6
discovery of antielectron and, 106–8,
183
Kaluza-Klein theory and, 148–49
matrices and, 103, 104, 117–18, 128,
160
unification of gravity and, 132, 184,
190–91
wave functions and, 103, 104–5, 111
quantum theory, 101, 102–4, 135, 137,
150, 154
quantum uncertainty, 167, 168
quarks, 114–15, 116, 136, 150, 155
charm and strange, 192
top and bottom, 192
up and down, 192
qubits, 141, 167, 168

radioactivity, 97, 102
Rees, Martin, 187, 188
Just Six Numbers, 186
reflection of light, 66
refraction of light, 66, 82
relativity, 99, 122, 128, 138
Dirac's electron equation and, 105,
124–27

relativity *(continued)*
  general, 129, 132, 137, 138, 147, 154, 160,
    179
  special, 137
revolutionary science, Kuhn's notion of,
  137–38
Ricci, Ostilio, 7–8
Roman Empire, 34, 35
Russell, Bertrand, 37, 46, 119–20, 124,
  127, 192–93

Santbech, Daniel, 5, 53
Saturn, 14, 20
Savart, Félix, 91
scalars, 177
Scholastics, 37
Schrödinger, Erwin, 103, 167
Schwartz, John, 133
scientific method, 44, 132
scientific theories, xiii
  experiment and observation as acid
    test of, 196
  falsifiability and legitimacy of, 189
  predictions made by, 190
"scientists," invention of word, 71
second industrial revolution, 169–70,
  171
Segrè, Emilio, 93–94
selectrons, 155
Shannon, Claude, 166
Shapin, Steven, *The Scientific
  Revolution,* 57–58
Simplicius, 30
sine curves, 75–79, 80, 82
  calculation of, 76–77
Smolin, Lee, 199–200
  Three Roads to Quantum Gravity,
    159, 194
  The Trouble with Physics, 158–59, 160,
    199
Socrates, 30, 52, 54
sound waves, 82

space-time:
  entropy and, 165
  extra dimensions of, 138, 147
  geometrical curvature of, 137, 147,
    160
  special relativity, 137
spin (quantum property), 105–6, 112–13
  Dirac's electron equation and, 105,
    124–27
squarks, 155
stars, 4, 15–16, 164
  supernovas, 20–21
steam engines, 84–85, 162
Steinhardt, Paul, 189
Stewart, Ian, 155
  Why Beauty Is Truth, 173
strangeness (quantum property), 113–14
strange quark, 192
string theory, 133, 136, 138, 155–61, 166,
  184–85, 191, 193–94, 198
  critics of, 158–59, 199–200
  mathematical erudition and, 159–61
  superstrings and, 155–57, 177, 179,
    190–91
strong nuclear force, 111, 149, 150, 154,
  192
  unification of electroweak force and,
    151–52
sun, 4, 8, 13, 14, 29
  believed to circle around earth, 9,
    15–16, 31
  at center of universe, 6–7, 11, 14–20;
    see also heliocentrism
  harmony of the spheres and, 28
SU(3) operations, 114–15
superconductivity, 140–41
supergravity, 155, 156
supernovas, 20–21
superpositions, 167, 168
superstring theories, 155–57, 177, 179
  extra dimensions in, 175, 191–92
supersymmetry, 154–55

Susskind, Leonard, 159, 193–94
symmetries, 157, 192, 198
    broken symmetries and, 149
    Higgs mechanism and, 149–50,
        152–53
    mathematics of, 112–16, 118–19
    supersymmetry and, 154–55
    Wigner's theorem and, 112–14

Tait, P. G., 194
tau and tau neutrino, 192
techne-episteme distinction, 142–43
Tegmark, Max:
    multiverse hypothesis and, 187, 189
    Our Mathematical Universe, 173–81
telescope, 66, 82
    Galileo's use of, 21–23, 41–42
    Hubble, 130
tensors, 178–79
theoretical physics, 102, 107–8, 115, 177
    beauty of mathematics and, 122–33
    science-engineering divide and,
        134–35
    see also particle physics; quantum
        theory; string theory
theory of everything, xi–xii, 132, 156–58,
    184–85, 190–91, 192–93, 195
thermal equilibrium, 163–64, 165
thermodynamics, 85–88, 135, 138, 148,
    175
    first law of, 86, 97, 102, 162
    second law of, 85, 97, 162, 165
thermometers, 83–84
third industrial revolution, 171
Thomism, 37, 47
Thomson, J. J., 103, 123
thought experiments:
    ascribed to Galileo, 55
    Einstein's use of, 55
tides, daily, 66, 190
Tridentine Creed and Mass, 37
trigonometry, 3–4, 76, 77, 78

uncertainty principle, 103
unification, 147–61
    of electromagnetism and gravity, 147,
        148–49
    of electroweak and strong forces,
        151–52
    grand unified theories, 151–54
    inflationary universe and, 152–54,
        189–90
    theory of everything, xi–xii, 132,
        156–58, 184–85, 190–91, 192–93,
        195
    of weak and electromagnetic forces,
        150–51
    see also string theory
universe:
    criteria for plausibility in thinking
        about, 170–71
    expansion of, 129–30, 165
    Galileo's dictum on mathematics as
        language of, 47–48, 54, 55–56, 81,
        120, 175
    life-supporting, conditions for,
        186–87, 188–89, 191–92
    machine metaphor for, 88–89, 95,
        96–97, 170, 171, 183
    multiverse hypothesis and, 158,
        187–93
    observable, 165
    seen as mathematics, 173, 174–81, 195
    seen as quantum computer, 168–71
    see also cosmology
upsilon, 115
Uranus, 72
Urban VIII, Pope (formerly Cardinal
    Maffeo Barberini), 45–46

vectors, 177
velocity vectors, 177
Venus, 14, 20
    phases and brightness of, 21–22,
        23, 43

violin string, vibration of, 76, 78, 79
Voltaire, 90
vortex atom theory, 123, 127

wave equation:
 light and, 82–83
 sine curves and, 75–79, 80, 82
wave functions:
 Dirac's electron equation and, 105,
  124–27
 particles described in terms of, 103,
  104–5, 111
weak nuclear force, 102, 149, 150,
 154
 unification of electromagnetic
  force and, 150–51
Weinberg, Steven, 20, 131, 154, 188
Weyl, Hermann, 122, 129

Wigner, Eugene, 110, 112–13, 129, 195
 use of pure mathematics in physics
  questioned by, 116–21, 173, 195
Witten, Edward, 156, 193
Wittgenstein, Ludwig, 174
Woit, Peter, *Not Even Wrong,* 199
Wootton, David, *The Invention of
 Science,* 58
W particle, 150

Yukawa, Hideki, 110–11

Zarlino, Gioseffo, 50, 51, 52
Zeno's paradox, 59, 60–61, 65
Zermelo, Ernst, 164
zero, concept of, 34
zoology, 141–42
Z particle, 150

David Lindley holds a PhD in astrophysics from the University of Sussex and has been an editor at *Nature, Science,* and *Science News.* Now a full-time writer, he is the author of *Uncertainty, The End of Physics, Where Does the Weirdness Go?, The Science of "Jurassic Park," Boltzmann's Atom,* and *Degrees Kelvin.* He was also the recipient of the Phi Beta Kappa science writing prize. He lives in Alexandria, Virginia.